Basic Electronics

Basic Electronics

Edited by
Christopher Gunn

Larsen & Keller
www.larsen-keller.com

Basic Electronics
Edited by Christopher Gunn
ISBN: 978-1-63549-686-4 (Hardback)

⊟ Larsen & Keller

Published by Larsen and Keller Education,
5 Penn Plaza,
19th Floor,
New York, NY 10001, USA

Cataloging-in-Publication Data

Basic electronics / edited by Christopher Gunn.
    p. cm.
Includes bibliographical references and index.
ISBN 978-1-63549-686-4
1. Electronics. 2. Electrical engineering. I. Gunn, Christopher.
TK7803 .B37 2018
621.381--dc23

For more information regarding Larsen and Keller Education and its products, please visit the publisher's website www.larsen-keller.com

Table of Contents

Preface

Electronics deals with the science of electrons and controlling electric energy. It is mainly concerned with designing and constructing circuits by using electric machines like transistors. It includes the study of characteristics and behavior of electrons in vacuum, semiconductors, gas and conductors. Electronics has many branches namely circuit design, analogue electronics, embedded systems, microelectronics, digital electronics, etc. This book elucidates the concepts and innovative models around prospective developments with respect to electronics. It is a valuable compilation of topics, ranging from the basic to the most complex theories and principles in this field. This textbook is an essential guide for both academicians and those who wish to pursue this discipline further.

A detailed account of the significant topics covered in this book is provided below:

Chapter 1- Electronics is a branch of physics and electrical engineering that studies to control electricity by regulating the flow of electrons. The chapter also talks briefly about semiconductors, which are elements that are neither good conductors nor bad conductors of electricity. These conducting properties can be altered using impurities which form the basis of diodes. This chapter will provide an integrated understanding of electronics and semiconductors.

Chapter 2- An assimilation of various electronic components such as diodes, transistors, resistors, capacitors, and inductors that are connected with a copper wire form a complete circuit. It allows for the flow of electricity and is known as an electronic circuit. Apart from the common function of a diode to allow electricity to pass through one end and resist through other, they can be crafted to perform other special functions. By changing the semiconductor element as well as the impurity, special-purpose diodes could be manufactured. The following chapter elucidates the various techniques related to electronic circuits.

Chapter 3- A transistor is a semiconductor that is used to increase and switch electronic power and signal. It is usually made up of silicon or germanium and has three separate terminals. In this section, two major types of transistors are discussed. They are bipolar junction transistors and unijunction transistors. The chapter serves as a source to understand the major categories related to transistors.

Chapter 4- A field-effect transistor (FET) is a type of transistor that controls the electrical behavior of a device through the use of an electric field. Field-effect transistors have an active channel through which charge carriers travel. Two types of FETs discussed in this chapter are junction field-effect transistor and metal oxide semiconductor field-effect transistor. This chapter is an overview of the subject matter incorporating all the major aspects of field-effect transistor.

It gives me an immense pleasure to thank our entire team for their efforts. Finally in the end, I would like to thank my family and colleagues who have been a great source of inspiration and support.

Editor

An Overview of Electronics

Electronics is a branch of physics and electrical engineering that studies to control electricity by regulating the flow of electrons. The chapter also talks briefly about semiconductors, which are elements that are neither good conductors nor bad conductors of electricity. These conducting properties can be altered using impurities which form the basis of diodes. This chapter will provide an integrated understanding of electronics and semiconductors.

Electronics

Electronics is the science of controlling electrical energy electrically, in which the electrons have a fundamental role. Electronics deals with electrical circuits that involve active electrical components such as vacuum tubes, transistors, diodes, integrated circuits, associated passive electrical components, and interconnection technologies. Commonly, electronic devices contain circuitry consisting primarily or exclusively of active semiconductors supplemented with passive elements; such a circuit is described as an electronic circuit.

Surface-mount electronic components

The science of electronics is also considered to be a branch of physics and electrical engineering.

The nonlinear behaviour of active components and their ability to control electron flows makes amplification of weak signals possible, and electronics is widely used in information processing, telecommunication, and signal processing. The ability of electronic devices to act as switches makes digital information processing possible. Interconnection technologies such as circuit boards, electronics packaging technology, and other varied forms of communication infrastructure complete circuit functionality and transform the mixed components into a regular working system.

Electronics is distinct from electrical and electro-mechanical science and technology, which deal with the generation, distribution, switching, storage, and conversion of electrical energy to and from other energy forms using wires, motors, generators, batteries, switches, relays, transformers, resistors, and other passive components. This distinction started around 1906 with the invention by Lee De Forest of the triode, which made electrical amplification of weak radio signals and audio signals possible with a non-mechanical device. Until 1950 this field was called "radio technology" because its principal application was the design and theory of radio transmitters, receivers, and vacuum tubes.

Today, most electronic devices use semiconductor components to perform electron control. The study of semiconductor devices and related technology is considered a branch of solid-state physics, whereas the design and construction of electronic circuits to solve practical problems come under electronics engineering.

Branches of Electronics

Electronics has branches as follows:

1. Digital electronics
2. Analogue electronics
3. Microelectronics
4. Circuit design
5. Integrated circuits
6. Optoelectronics
7. Semiconductor devices
8. Embedded systems

Electronic Devices and Components

An electronic component is any physical entity in an electronic system used to affect the electrons or their associated fields in a manner consistent with the intended function of the electronic system. Components are generally intended to be connected together, usually by being soldered to a printed circuit board (PCB), to create an electronic circuit with a particular function (for example an amplifier, radio receiver, or oscillator). Components may be packaged singly, or in more complex groups as integrated circuits. Some common electronic components are capacitors, inductors, resistors, diodes, transistors, etc. Components are often categorized as active (e.g. transistors and thyristors) or passive (e.g. resistors, diodes, inductors and capacitors).

History of Electronic Components

Vacuum tubes (Thermionic valves) were among the earliest electronic components. They were almost solely responsible for the electronics revolution of the first half of the Twentieth Century. They took electronics from parlor tricks and gave us radio, television, phonographs, radar, long distance telephony and much more. They played a leading role in the field of microwave and high power transmission as well as television receivers until the middle of the 1980s. Since that time, solid state devices have all but completely taken over. Vacuum tubes are still used in some specialist applications such as high power RF amplifiers, cathode ray tubes, specialist audio equipment, guitar amplifiers and some microwave devices.

In April 1955 the IBM 608 was the first IBM product to use transistor circuits without any vacuum tubes and is believed to be the world's first all-transistorized calculator to be manufactured for the commercial market. The 608 contained more than 3,000 germanium transistors. Thomas J. Watson Jr. ordered all future IBM products to use transistors in their design. From that time on transistors were almost exclusively used for computer logic and peripherals.

Types of Circuits

Circuits and components can be divided into two groups: analog and digital. A particular device may consist of circuitry that has one or the other or a mix of the two types.

Analog Circuits

Most analog electronic appliances, such as radio receivers, are constructed from combinations of a few types of basic circuits. Analog circuits use a continuous range of voltage or current as opposed to discrete levels as in digital circuits.

The number of different analog circuits so far devised is huge, especially because a 'circuit' can be defined as anything from a single component, to systems containing thousands of components.

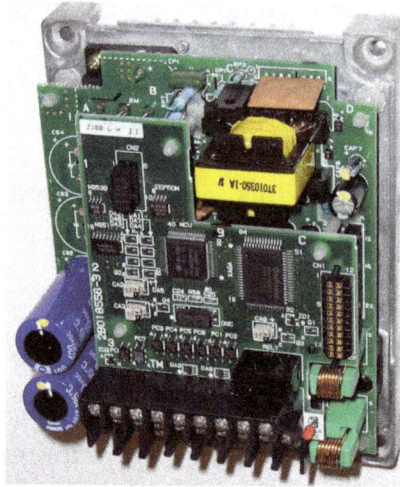

Hitachi J100 adjustable frequency drive chassis

Analog circuits are sometimes called linear circuits although many non-linear effects are used in analog circuits such as mixers, modulators, etc. Good examples of analog circuits include vacuum tube and transistor amplifiers, operational amplifiers and oscillators.

One rarely finds modern circuits that are entirely analog. These days analog circuitry may use digital or even microprocessor techniques to improve performance. This type of circuit is usually called "mixed signal" rather than analog or digital.

Sometimes it may be difficult to differentiate between analog and digital circuits as they have elements of both linear and non-linear operation. An example is the comparator which takes in a continuous range of voltage but only outputs one of two levels as in a digital circuit. Similarly, an overdriven transistor amplifier can take on the characteristics of a controlled switch having essentially two levels of output. In fact, many digital circuits are actually implemented as variations of analog circuits similar to this example—after all, all aspects of the real physical world are essentially analog, so digital effects are only realized by constraining analog behavior.

Digital Circuits

Digital circuits are electric circuits based on a number of discrete voltage levels. Digital circuits are the most common physical representation of Boolean algebra, and are the basis of all digital computers. To most engineers, the terms "digital circuit", "digital system" and "logic" are interchangeable in the context of digital circuits. Most digital circuits use a binary system with two voltage levels labeled "0" and "1". Often logic "0" will be a lower voltage and referred to as "Low" while logic "1" is referred to as "High". However, some systems use the reverse definition ("0" is "High") or are current based. Quite often the logic designer may reverse these definitions from one circuit to the next as he sees fit to facilitate his design. The definition of the levels as "0" or "1" is arbitrary.

Ternary (with three states) logic has been studied, and some prototype computers made.

Computers, electronic clocks, and programmable logic controllers (used to control industrial processes) are constructed of digital circuits. Digital signal processors are another example.

Building blocks:

- Logic gates

- Adders

- Flip-flops

- Counters

- Registers

- Multiplexers

- Schmitt triggers

Highly integrated devices:

- Microprocessors

- Microcontrollers

- Application-specific integrated circuit (ASIC)

- Digital signal processor (DSP)

- Field-programmable gate array (FPGA)

Heat Dissipation and Thermal Management

Heat generated by electronic circuitry must be dissipated to prevent immediate failure and improve long term reliability. Heat dissipation is mostly achieved by passive conduction/convection. Means to achieve greater dissipation include heat sinks and fans for air cooling, and other forms of computer cooling such as water cooling. These techniques use convection, conduction, and radiation of heat energy.

Noise

Electronic noise is defined as unwanted disturbances superposed on a useful signal that tend to obscure its information content. Noise is not the same as signal distortion caused by a circuit. Noise is associated with all electronic circuits. Noise may be electromagnetically or thermally generated, which can be decreased by lowering the operating temperature of the circuit. Other types of noise, such as shot noise cannot be removed as they are due to limitations in physical properties.

Electronics Theory

Mathematical methods are integral to the study of electronics. To become proficient in electronics it is also necessary to become proficient in the mathematics of circuit analysis.

Circuit analysis is the study of methods of solving generally linear systems for unknown variables such as the voltage at a certain node or the current through a certain branch of a network. A common analytical tool for this is the SPICE circuit simulator.

Also important to electronics is the study and understanding of electromagnetic field theory.

Electronics Lab

Due to the complex nature of electronics theory, laboratory experimentation is an important part of the development of electronic devices. These experiments are used to test or verify the engineer's design and detect errors. Historically, electronics labs have consisted of electronics devices and equipment located in a physical space, although in more recent years the trend has been towards electronics lab simulation software, such as CircuitLogix, Multisim, and PSpice.

Computer Aided Design (CAD)

Today's electronics engineers have the ability to design circuits using premanufactured building blocks such as power supplies, semiconductors (i.e. semiconductor devices, such as transistors), and integrated circuits. Electronic design automation software programs include schematic capture programs and printed circuit board design programs. Popular names in the EDA software world are NI Multisim, Cadence (ORCAD), EAGLE PCB and Schematic, Mentor (PADS PCB and LOGIC Schematic), Altium (Protel), LabCentre Electronics (Proteus), gEDA, KiCad and many others.

Construction Methods

Many different methods of connecting components have been used over the years. For instance, early electronics often used point to point wiring with components attached to wooden breadboards to construct circuits. Cordwood construction and wire wrap were other methods used. Most modern day electronics now use printed circuit boards made of materials such as FR4, or the cheaper (and less hard-wearing) Synthetic Resin Bonded Paper (SRBP, also known as Paxoline/Paxolin (trade marks) and

FR2) - characterised by its brown colour. Health and environmental concerns associated with electronics assembly have gained increased attention in recent years, especially for products destined to the European Union, with its Restriction of Hazardous Substances Directive (RoHS) and Waste Electrical and Electronic Equipment Directive (WEEE), which went into force in July 2006.

Semiconductor

Semiconductors are crystalline or amorphous solids with distinct electrical characteristics. They are of high electrical resistance — higher than typical resistance materials, but still of much lower resistance than insulators. Their resistance decreases as their temperature increases, which is behavior opposite to that of a metal. Their conducting properties may be altered in useful ways by the deliberate, controlled introduction of impurities ("doping") into the crystal structure, which lowers its resistance but also permits the creation of semiconductor junctions between differently-doped regions of the extrinsic semiconductor crystal. The behavior of charge carriers which include electrons, ions and electron holes at these junctions is the basis of diodes, transistors and all modern electronics.

Semiconductor devices can display a range of useful properties such as passing current more easily in one direction than the other, showing variable resistance, and sensitivity to light or heat. Because the electrical properties of a semiconductor material can be modified by doping, or by the application of electrical fields or light, devices made from semiconductors can be used for amplification, switching, and energy conversion.

The modern understanding of the properties of a semiconductor relies on quantum physics to explain the movement of charge carriers in a crystal lattice. Doping greatly increases the number of charge carriers within the crystal. When a doped semiconductor contains mostly free holes it is called "p-type", and when it contains mostly free electrons it is known as "n-type". The semiconductor materials used in electronic devices are doped under precise conditions to control the concentration and regions of p- and n-type dopants. A single semiconductor crystal can have many p- and n-type regions; the p–n junctions between these regions are responsible for the useful electronic behaviour.

Although some pure elements and many compounds display semiconductor properties, silicon, germanium, and compounds of gallium are the most widely used in electronic devices. Elements near the so-called "metalloid staircase", where the metalloids are located on the periodic table, are usually used as semiconductors.

Some of the properties of semiconductor materials were observed throughout the mid 19th and first decades of the 20th century. The first practical application of semiconductors in electronics was the 1904 development of the Cat's-whisker detector, a primitive semiconductor diode widely used in early radio receivers. Developments in quantum physics in turn allowed the development of the transistor in 1947 and the integrated circuit in 1958.

Properties

Variable conductivity

Semiconductors in their natural state are poor conductors because a current requires the flow of electrons, and semiconductors have their valence bands filled, preventing the entry flow of new electrons. There are several developed techniques that allow semiconducting materials to behave like conducting materials, such as doping or gating. These modifications have two outcomes: n-type and p-type. These refer to the excess or shortage of electrons, respectively. An unbalanced number of electrons would cause a current to flow through the material.

Heterojunctions

Heterojunctions occur when two differently doped semiconducting materials are joined together. For example, a configuration could consist of p-doped and n-doped germanium. This results in an exchange of electrons and holes between the differently doped semiconducting materials. The n-doped germanium would have an excess of electrons, and the p-doped germanium would have an excess of holes. The transfer occurs until equilibrium is reached by a process called recombination, which causes the migrating electrons from the n-type to come in contact with the migrating holes from the p-type. A product of this process is charged ions, which result in an electric field.

Excited Electrons

A difference in electric potential on a semiconducting material would cause it to leave thermal equilibrium and create a non-equilibrium situation. This introduces electrons and holes to the system, which interact via a process called ambipolar diffusion. Whenever thermal equilibrium is disturbed in a semiconducting material, the amount of holes and electrons changes. Such disruptions can occur as a result of a temperature difference or photons, which can enter the system and create electrons and holes. The process that creates and annihilates electrons and holes are called generation and recombination.

Light emission

In certain semiconductors, excited electrons can relax by emitting light instead of producing heat. These semiconductors are used in the construction of light-emitting diodes and fluorescent quantum dots.

Thermal energy conversion

> Semiconductors have large thermoelectric power factors making them useful in thermoelectric generators, as well as high thermoelectric figures of merit making them useful in thermoelectric coolers.

Materials

Silicon crystals are the most common semiconducting materials used in microelectronics and photovoltaics.

A large number of elements and compounds have semiconducting properties, including:

- Certain pure elements are found in Group 14 of the periodic table; the most commercially important of these elements are silicon and germanium. Silicon and germanium are used here effectively because they have 4 valence electrons in their outermost shell which gives them the ability to gain or lose electrons equally at the same time.

- Binary compounds, particularly between elements in Groups 13 and 15, such as gallium arsenide, Groups 12 and 16, groups 14 and 16, and between different group 14 elements, e.g. silicon carbide.

- Certain ternary compounds, oxides and alloys.

- Organic semiconductors, made of organic compounds.

Most common semiconducting materials are crystalline solids, but amorphous and liquid semiconductors are also known. These include hydrogenated amorphous silicon and mixtures of arsenic, selenium and tellurium in a variety of proportions. These compounds share with better known semiconductors the properties of intermediate conductivity and a rapid variation of conductivity with temperature, as well as occasional negative resistance. Such disordered materials lack the rigid crystalline structure of conventional semiconductors such as silicon. They are generally used in thin film structures, which do not require material of higher electronic quality, being relatively insensitive to impurities and radiation damage.

Preparation of Semiconductor Materials

Almost all of today's electronic technology involves the use of semiconductors, with the most important aspect being the integrated circuit (IC), which are found in laptops, scanners, cell-phones, etc. Semiconductors for ICs are mass-produced. To create an ideal semiconducting material, chemical purity is paramount. Any small imperfection can have a drastic effect on how the semiconducting material behaves due to the scale at which the materials are used.

A high degree of crystalline perfection is also required, since faults in crystal structure (such as dislocations, twins, and stacking faults) interfere with the semiconducting properties of the material. Crystalline faults are a major cause of defective semiconductor devices. The larger the crystal, the more difficult it is to achieve the necessary perfection. Current mass production processes use crystal ingots between 100 and 300 mm (4 and 12 in) in diameter which are grown as cylinders and sliced into wafers.

There is a combination of processes that is used to prepare semiconducting materials for ICs. One process is called thermal oxidation, which forms silicon dioxide on the surface of the silicon. This is used as a gate insulator and field oxide. Other processes are called photomasks and photolithography. This process is what creates the patterns on the circuity in the integrated circuit. Ultraviolet light is used along with a photoresist layer to create a chemical change that generates the patterns for the circuit.

Etching is the next process that is required. The part of the silicon that was not covered by the photoresist layer from the previous step can now be etched. The main process typically used today is called plasma etching. Plasma etching usually involves an etch gas pumped in a low-pressure chamber to create plasma. A common etch gas is chlorofluorocarbon, or more commonly known Freon. A high radio-frequency voltage between the cathode and anode is what creates the plasma in the chamber. The silicon wafer is located on the cathode, which causes it to be hit by the positively charged ions that are released from the plasma. The end result is silicon that is etched anisotropically.

The last process is called diffusion. This is the process that gives the semiconducting material its desired semiconducting properties. It is also known as doping. The process introduces an impure atom to the system, which creates the p-n junction. In order to get the impure atoms embedded in the silicon wafer, the wafer is first put in a 1100 degree Celsius chamber. The atoms are injected in and eventually diffuse with the silicon. After the process is completed and the silicon has reached room temperature, the doping process is done and the semiconducting material is ready to be used in an integrated circuit.

Physics of Semiconductors

Energy Bands and Electrical Conduction

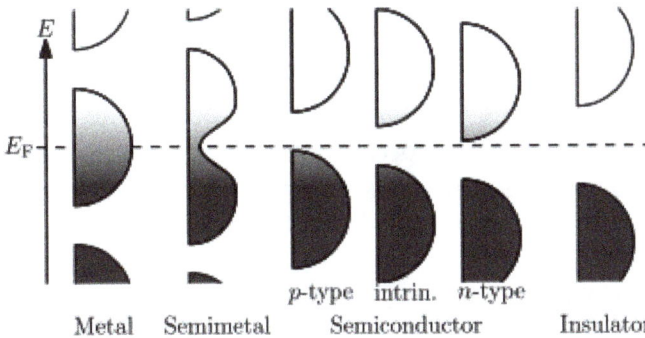

Filling of the electronic states in various types of materials at equilibrium. Here, height is energy while width is the density of available states for a certain energy in the material listed. The shade follows the Fermi–Dirac distribution (black = all states filled, white = no state filled). In metals and semimetals the Fermi level E_F lies inside at least one band. In insulators and semiconductors the Fermi level is inside a band gap; however, in semiconductors the bands are near enough to the Fermi level to be thermally populated with electrons or holes.

Semiconductors are defined by their unique electric conductive behavior, somewhere between that of a conductor and an insulator. The differences between these materials can be understood in terms of the quantum states for electrons, each of which may contain zero or one electron (by the Pauli exclusion principle). These states are associated with the electronic band structure of the material. Electrical conductivity arises due to the presence of electrons in states that are delocalized (extending through the material), however in order to transport electrons a state must be *partially filled*, containing an electron only part of the time. If the state is always occupied with an electron, then it is inert, blocking the passage of other electrons via that state. The energies of these quantum states are critical, since a state is partially filled only if its energy is near the Fermi level.

High conductivity in a material comes from it having many partially filled states and much state delocalization. Metals are good electrical conductors and have many partially filled states with energies near their Fermi level. Insulators, by contrast, have few partially filled states, their Fermi levels sit within band gaps with few energy states to occupy. Importantly, an insulator can be made to conduct by increasing its temperature: heating provides energy to promote some electrons across the band gap, inducing partially filled states in both the band of states beneath the band gap (valence band) and the band of states above the band gap (conduction band). An (intrinsic) semiconductor has a band gap that is smaller than that of an insulator and at room temperature significant numbers of electrons can be excited to cross the band gap.

A pure semiconductor, however, is not very useful, as it is neither a very good insulator nor a very good conductor. However, one important feature of semiconductors (and some insulators, known as *semi-insulators*) is that their conductivity can be increased and controlled by doping with impurities and gating with electric fields. Doping and gating move either the conduction or valence band much closer to the Fermi level, and greatly increase the number of partially filled states.

Some wider-band gap semiconductor materials are sometimes referred to as semi-insulators. When undoped, these have electrical conductivity nearer to that of electrical insulators, however they can be doped (making them as useful as semiconductors). Semi-insulators find niche applications in micro-electronics, such as substrates for HEMT. An example of a common semi-insulator is gallium arsenide. Some materials, such as titanium dioxide, can even be used as insulating materials for some applications, while being treated as wide-gap semiconductors for other applications.

Charge Carriers (Electrons and Holes)

The partial filling of the states at the bottom of the conduction band can be understood as adding electrons to that band. The electrons do not stay indefinitely (due to the natural thermal recombination) but they can move around for some time. The actual concentration of electrons is typically very dilute, and so (unlike in metals) it is possible to think of the electrons in the conduction band of a semiconductor as a sort of classical ideal gas, where the electrons fly around freely without being subject to the Pauli exclusion principle. In most semiconductors the conduction bands have a parabolic dispersion relation, and so these electrons respond to forces (electric field, magnetic field, etc.) much like they would in a vacuum, though with a different effective mass. Because the electrons behave like an ideal gas, one may also think about conduction in very simplistic terms such as the Drude model, and introduce concepts such as electron mobility.

For partial filling at the top of the valence band, it is helpful to introduce the concept of an electron hole. Although the electrons in the valence band are always moving around, a completely full valence band is inert, not conducting any current. If an electron is taken out of the valence band, then the trajectory that the electron would normally have taken is now missing its charge. For the purposes of electric current, this combination of the full valence band, minus the electron, can be converted into a picture of a completely empty band containing a positively charged particle that moves in the same way as the electron. Combined with the *negative* effective mass of the electrons at the top of the valence band, we arrive at a picture of a positively charged particle that responds to electric and magnetic fields just as a normal positively charged particle would do in vacuum, again with some positive effective mass. This particle is called a hole, and the collection of holes in the valence band can again be understood in simple classical terms (as with the electrons in the conduction band).

Carrier Generation and Recombination

When ionizing radiation strikes a semiconductor, it may excite an electron out of its energy level and consequently leave a hole. This process is known as *electron–hole pair generation*. Electron-hole pairs are constantly generated from thermal energy as well, in the absence of any external energy source.

Electron-hole pairs are also apt to recombine. Conservation of energy demands that these recombination events, in which an electron loses an amount of energy larger than the band gap, be accompanied by the emission of thermal energy (in the form of phonons) or radiation (in the form of photons).

In some states, the generation and recombination of electron–hole pairs are in equipoise. The number of electron-hole pairs in the steady state at a given temperature is determined by quantum statistical mechanics. The precise quantum mechanical mechanisms of generation and recombination are governed by conservation of energy and conservation of momentum.

As the probability that electrons and holes meet together is proportional to the product of their amounts, the product is in steady state nearly constant at a given temperature, providing that there is no significant electric field (which might "flush" carriers of both types, or move them from neighbour regions containing more of them to meet together) or externally driven pair generation. The product is a function of the temperature, as the probability of getting enough thermal energy to produce a pair increases with temperature, being approximately $\exp(-E_G/kT)$, where k is Boltzmann's constant, T is absolute temperature and E_G is band gap.

The probability of meeting is increased by carrier traps—impurities or dislocations which can trap an electron or hole and hold it until a pair is completed. Such carrier traps are sometimes purposely added to reduce the time needed to reach the steady state.

Doping

The conductivity of semiconductors may easily be modified by introducing impurities into their crystal lattice. The process of adding controlled impurities to a semiconductor is known as *doping*. The amount of impurity, or dopant, added to an *intrinsic* (pure) semiconductor varies its level of conductivity. Doped semiconductors are referred to as *extrinsic*. By adding impurity to the pure semiconductors, the electrical conductivity may be varied by factors of thousands or millions.

A 1 cm^3 specimen of a metal or semiconductor has of the order of 10^{22} atoms. In a metal, every atom donates at least one free electron for conduction, thus 1 cm^3 of metal contains on the order of 10^{22} free electrons, whereas a 1 cm^3 sample of pure germanium at 20 °C contains about 4.2×10^{22} atoms, but only 2.5×10^{13} free electrons and 2.5×10^{13} holes. The addition of 0.001% of arsenic (an impurity) donates an extra 10^{17} free electrons in the same volume and the electrical conductivity is increased by a factor of 10,000.

The materials chosen as suitable dopants depend on the atomic properties of both the dopant and the material to be doped. In general, dopants that produce the desired controlled changes are classified as either electron acceptors or donors. Semiconductors doped with *donor* impurities are called *n-type*, while those doped with *acceptor* impurities are known as *p-type*. The n and p type designations indicate which charge carrier acts as the material's majority carrier. The opposite carrier is called the minority carrier, which exists due to thermal excitation at a much lower concentration compared to the majority carrier.

For example, the pure semiconductor silicon has four valence electrons which bond each silicon atom to its neighbors. In silicon, the most common dopants are *group III* and *group V* elements. Group III elements all contain three valence electrons, causing them to function as acceptors when used to dope silicon. When an acceptor atom replaces a silicon atom in the crystal, a vacant state (an electron "hole") is created, which can move around the lattice and functions as a charge carrier. Group V elements have five valence electrons, which allows them to act as a donor; substitution of these atoms for silicon creates an extra free electron. Therefore, a silicon crystal doped with boron creates a p-type semiconductor whereas one doped with phosphorus results in an n-type material.

During manufacture, dopants can be diffused into the semiconductor body by contact with gaseous compounds of the desired element, or ion implantation can be used to accurately position the doped regions.

Early History of Semiconductors

The history of the understanding of semiconductors begins with experiments on the electrical properties of materials. The properties of negative temperature coefficient of resistance, rectification, and light-sensitivity were observed starting in the early 19th century.

Thomas Johann Seebeck was the first to notice an effect due to semiconductors, in 1821. In 1833, Michael Faraday reported that the resistance of specimens of silver sulfide decreases when they are heated. This is contrary to the behavior of metallic substances such as copper. In 1839, A. E. Becquerel reported observation of a voltage between a solid and a liquid electrolyte when struck by light, the photovoltaic effect. In 1873 Willoughby Smith observed that selenium resistors exhibit decreasing resistance when light falls on them. In 1874 Karl Ferdinand Braun observed conduction and rectification in metallic sulphides, although this effect had been discovered much earlier by M.A. Rosenschold writing for the Annalen der Physik und Chemie in 1835, and Arthur Schuster found that a copper oxide layer on wires has rectification properties that ceases when the wires are cleaned. Adams and Day observed the photovoltaic effect in selenium in 1876.

A unified explanation of these phenomena required a theory of solid-state physics which developed greatly in the first half of the 20th Century. In 1878 Edwin Herbert

Hall demonstrated the deflection of flowing charge carriers by an applied magnetic field, the Hall effect. The discovery of the electron by J.J. Thomson in 1897 prompted theories of electron-based conduction in solids. Karl Baedeker, by observing a Hall effect with the reverse sign to that in metals, theorized that copper iodide had positive charge carriers. Johan Koenigsberger classified solid materials as metals, insulators and "variable conductors" in 1914 although his student Josef Weiss already introduced the term *Halbleiter* (semiconductor in modern meaning) in PhD thesis in 1910. Felix Bloch published a theory of the movement of electrons through atomic lattices in 1928. In 1930, B. Gudden stated that conductivity in semiconductors was due to minor concentrations of impurities. By 1931, the band theory of conduction had been established by Alan Herries Wilson and the concept of band gaps had been developed. Walter H. Schottky and Nevill Francis Mott developed models of the potential barrier and of the characteristics of a metal-semiconductor junction. By 1938, Boris Davydov had developed a theory of the copper-oxide rectifier, identifying the effect of the p–n junction and the importance of minority carriers and surface states.

Agreement between theoretical predictions (based on developing quantum mechanics) and experimental results was sometimes poor. This was later explained by John Bardeen as due to the extreme "structure sensitive" behavior of semiconductors, whose properties change dramatically based on tiny amounts of impurities. Commercially pure materials of the 1920s containing varying proportions of trace contaminants produced differing experimental results. This spurred the development of improved material refining techniques, culminating in modern semiconductor refineries producing materials with parts-per-trillion purity.

Devices using semiconductors were at first constructed based on empirical knowledge, before semiconductor theory provided a guide to construction of more capable and reliable devices.

Alexander Graham Bell used the light-sensitive property of selenium to transmit sound over a beam of light in 1880. A working solar cell, of low efficiency, was constructed by Charles Fritts in 1883 using a metal plate coated with selenium and a thin layer of gold; the device became commercially useful in photographic light meters in the 1930s. Point-contact microwave detector rectifiers made of lead sulfide were used by Jagadish Chandra Bose in 1904; the cat's-whisker detector using natural galena or other materials became a common device in the development of radio. However, it was somewhat unpredictable in operation and required manual adjustment for best performance. In 1906 H.J. Round observed light emission when electric current passed through silicon carbide crystals, the principle behind the light-emitting diode. Oleg Losev observed similar light emission in 1922 but at the time the effect had no practical use. Power rectifiers, using copper oxide and selenium, were developed in the 1920s and became commercially important as an alternative to vacuum tube rectifiers.

In the years preceding World War II, infra-red detection and communications devices prompted research into lead-sulfide and lead-selenide materials. These devices were used for detecting ships and aircraft, for infrared rangefinders, and for voice communication systems. The point-contact crystal detector became vital for microwave radio systems, since available vacuum tube devices could not serve as detectors above about 4000 MHz; advanced radar systems relied on the fast response of crystal detectors. Considerable research and development of silicon materials occurred during the war to develop detectors of consistent quality.

Detector and power rectifiers could not amplify a signal. Many efforts were made to develop a solid-state amplifier, but these were unsuccessful because of limited theoretical understanding of semiconductor materials. In 1922 Oleg Losev developed two-terminal, negative resistance amplifiers for radio; however, he perished in the Siege of Leningrad. In 1926 Julius Edgar Lilienfeld patented a device resembling a modern field-effect transistor, but it was not practical. R. Hilsch and R. W. Pohl in 1938 demonstrated a solid-state amplifier using a structure resembling the control grid of a vacuum tube; although the device displayed power gain, it had a cut-off frequency of one cycle per second, too low for any practical applications, but an effective application of the available theory. At Bell Labs, William Shockley and A. Holden started investigating solid-state amplifiers in 1938. The first p–n junction in silicon was observed by Russell Ohl about 1941, when a specimen was found to be light-sensitive, with a sharp boundary between p-type impurity at one end and n-type at the other. A slice cut from the specimen at the p–n boundary developed a voltage when exposed to light.

In France, during the war, Herbert Mataré had observed amplification between adjacent point contacts on a germanium base. After the war, Mataré's group announced their "Transistron" amplifier only shortly after Bell Labs announced the "transistor".

Diode

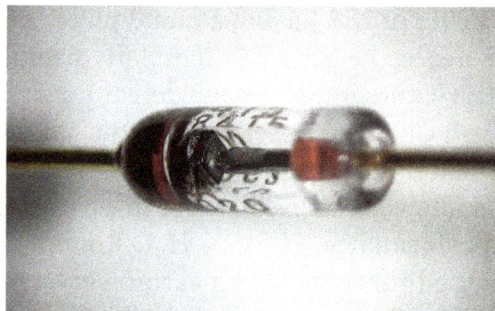

Closeup of a diode, showing the square-shaped semiconductor crystal *(black object on left)*.

In electronics, a diode is a two-terminal electronic component that conducts primarily in one direction (asymmetric conductance); it has low (ideally zero) resistance to the current in one direction, and high (ideally infinite) resistance in the other. A semiconductor diode, the most common type today, is a crystalline piece of semiconductor material with a p–n junction connected to two electrical terminals. A vacuum tube diode has two electrodes, a plate (anode) and a heated cathode. Semiconductor diodes were the first semiconductor electronic devices. The discovery of crystals' rectifying abilities was made by German physicist Ferdinand Braun in 1874. The first semiconductor diodes, called cat's whisker diodes, developed around 1906, were made of mineral crystals such as galena. Today, most diodes are made of silicon, but other semiconductors such as selenium and germanium are sometimes used.

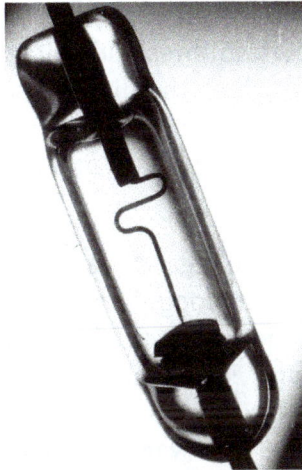

Extreme macro photo of a Chinese diode of the seventies.

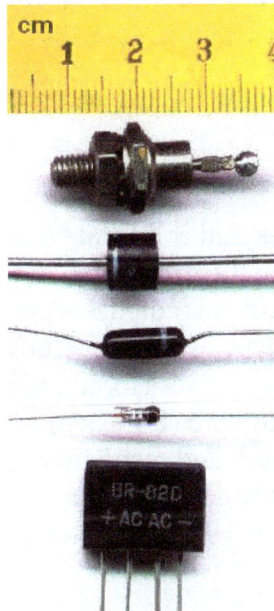

Various semiconductor diodes. Bottom: A bridge rectifier. In most diodes, a white or black painted band identifies the cathode into which electrons will flow when the diode is conducting. Electron flow is the reverse of conventional current flow.

Main Functions

The most common function of a diode is to allow an electric current to pass in one direction (called the diode's *forward* direction), while blocking current in the opposite direction (the *reverse* direction). Thus, the diode can be viewed as an electronic version of a check valve. This unidirectional behavior is called rectification, and is used to convert alternating current (AC) to direct current (DC), including extraction of modulation from radio signals in radio receivers—these diodes are forms of rectifiers.

However, diodes can have more complicated behavior than this simple on–off action, because of their nonlinear current-voltage characteristics. Semiconductor diodes begin conducting electricity only if a certain threshold voltage or cut-in voltage is present in the forward direction (a state in which the diode is said to be *forward-biased*). The voltage drop across a forward-biased diode varies only a little with the current, and is a function of temperature; this effect can be used as a temperature sensor or as a voltage reference.

A semiconductor diode's current–voltage characteristic can be tailored by selecting the semiconductor materials and the doping impurities introduced into the materials during manufacture. These techniques are used to create special-purpose diodes that perform many different functions. For example, diodes are used to regulate voltage (Zener diodes), to protect circuits from high voltage surges (avalanche diodes), to electronically tune radio and TV receivers (varactor diodes), to generate radio-frequency oscillations (tunnel diodes, Gunn diodes, IMPATT diodes), and to produce light (light-emitting diodes). Tunnel, Gunn and IMPATT diodes exhibit negative resistance, which is useful in microwave and switching circuits.

Diodes, both vacuum and semiconductor, can be used as shot-noise generators.

History

Thermionic (vacuum tube) diodes and solid state (semiconductor) diodes were developed separately, at approximately the same time, in the early 1900s, as radio receiver detectors. Until the 1950s vacuum tube diodes were used more frequently in radios because the early point-contact type semiconductor diodes were less stable. In addition, most receiving sets had vacuum tubes for amplification that could easily have the thermionic diodes included in the tube (for example the 12SQ7 double diode triode), and vacuum tube rectifiers and gas-filled rectifiers were capable of handling some high voltage/high current rectification tasks better than the semiconductor diodes (such as selenium rectifiers) which were available at that time.

Vacuum Tube Diodes

In 1873, Frederick Guthrie discovered the basic principle of operation of thermionic diodes. Guthrie discovered that a positively charged electroscope could be discharged by bringing a grounded piece of white-hot metal close to it (but not actually touching it). The same did not apply to a negatively charged electroscope, indicating that the current flow was only possible in one direction.

Thomas Edison independently rediscovered the principle on February 13, 1880. At the time, Edison was investigating why the filaments of his carbon-filament light bulbs nearly always burned out at the positive-connected end. He had a special bulb made with a metal plate sealed into the glass envelope. Using this device, he confirmed that an invisible current flowed from the glowing filament through the vacuum to the metal plate, but only when the plate was connected to the positive supply.

Edison devised a circuit where his modified light bulb effectively replaced the resistor in a DC voltmeter. Edison was awarded a patent for this invention in 1884. Since there was no apparent practical use for such a device at the time, the patent application was most likely simply a precaution in case someone else did find a use for the so-called Edison effect.

About 20 years later, John Ambrose Fleming (scientific adviser to the Marconi Company and former Edison employee) realized that the Edison effect could be used as a precision radio detector. Fleming patented the first true thermionic diode, the Fleming valve, in Britain on November 16, 1904 (followed by U.S. Patent 803,684 in November 1905).

Solid-state Diodes

In 1874 German scientist Karl Ferdinand Braun discovered the "unilateral conduction" of crystals. Braun patented the crystal rectifier in 1899. Copper oxide and selenium rectifiers were developed for power applications in the 1930s.

Indian scientist Jagadish Chandra Bose was the first to use a crystal for detecting radio waves in 1894. The crystal detector was developed into a practical device for wireless telegraphy by Greenleaf Whittier Pickard, who invented a silicon crystal detector in 1903 and received a patent for it on November 20, 1906. Other experimenters tried a variety of other substances, of which the most widely used was the mineral galena (lead sulfide). Other substances offered slightly better performance, but galena was most widely used because it had the advantage of being cheap and easy to obtain. The crystal detector in these early crystal radio sets consisted of an adjustable wire point-contact, often made of gold or platinum because of their incorrodible nature (the so-called "cat's whisker"), which could be manually moved over the face of the crystal in search of a portion of that mineral with rectifying qualties. This troublesome device was superseded by thermionic diodes (vacuum tubes) by the 1920s, but after high purity semiconductor materials became available, the crystal detector returned to dominant use with the advent, in the 1950s, of inexpensive fixed-germanium diodes. Bell Labs also developed a germanium diode for microwave reception, and AT&T used these in their microwave towers that criss-crossed the nation starting in the late 1940s, carrying telephone and network television signals. Bell Labs did not develop a satisfactory thermionic diode for microwave reception.

Etymology

At the time of their invention, such devices were known as rectifiers. In 1919, the year tetrodes were invented, William Henry Eccles coined the term *diode* from the Greek roots *di* (meaning 'two', and *ode*), meaning 'path'. (However, the word *diode* itself, as well as *triode, tetrode, pentode, hexode*, were already in use as terms of multiplex telegraphy.

Rectifiers

Although all diodes *rectify*, the term 'rectifier' is normally reserved for higher currents and voltages than would normally be found in the rectification of lower power signals; examples include:

- Power supply rectifiers (*half-wave, full-wave, bridge*)
- Flyback diodes

World's Smallest Diode

Researchers from the University of Georgia and Ben-Gurion University of the Negev (BGU) have developed a diode made from a molecule of DNA. Professor Bingqian Xu from the College of Engineering at the University of Georgia and his team took a single DNA molecule made from 11 base pairs and connected it to an electronic circuit a few nanometers in size. When layers of coralyne were inserted between layers of DNA, the current jumped up to 15 times larger negative versus positive, which is necessary for a nano diode.

Thermionic Diodes

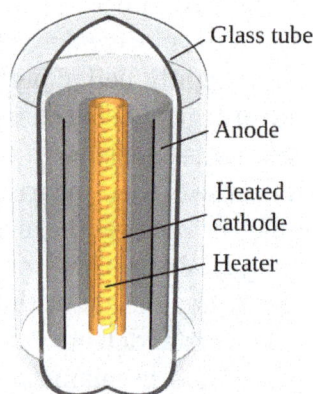

Diode vacuum tube construction

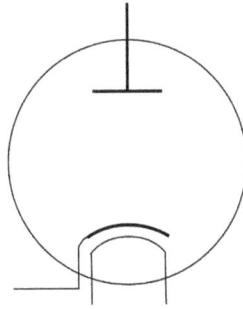

The symbol for an indirect heated vacuum-tube diode. From top to bottom, the components
are the anode, the cathode, and the heater filament.

A thermionic diode is a thermionic-valve device (also known as a vacuum tube, tube, or valve), consisting of a sealed evacuated glass envelope containing two electrodes: a cathode heated by a filament, and a plate (anode). Early examples were fairly similar in appearance to incandescent light bulbs.

In operation, a current flows through the filament (heater)—a high resistance wire made of nichrome—and heats the cathode red hot (800–1000 °C). This causes the cathode to release electrons into the vacuum, a process called thermionic emission. (Some valves use *direct heating*, in which a tungsten filament acts as both heater and cathode.) The alternating voltage to be rectified is applied between the cathode and the concentric plate electrode. When the plate has a positive voltage with respect to the cathode, it electrostatically attracts the electrons from the cathode, so a current of electrons flows through the tube from cathode to plate. However, when the polarity is reversed and the plate has a negative voltage, no current flows, because the cathode electrons are not attracted to it. The plate, being unheated, does not emit any electrons. So electrons can only flow through the tube in one direction, from the cathode to the anode plate.

The cathode is coated with oxides of alkaline earth metals, such as barium and strontium oxides. These have a low work function, meaning that they more readily emit electrons than would the uncoated cathode.

In a mercury-arc valve, an arc forms between a refractory conductive anode and a pool of liquid mercury acting as cathode. Such units were made with ratings up to hundreds of kilowatts, and were important in the development of HVDC power transmission. Some types of smaller thermionic rectifiers had mercury vapor fill to reduce their forward voltage drop and to increase current rating over thermionic hard-vacuum devices.

Throughout the vacuum tube era, valve diodes were used in analog signal applications and as rectifiers in DC power supplies in consumer electronics such as radios, televisions, and sound systems. They were replaced in power supplies beginning in the 1940s by selenium rectifiers and then by semiconductor diodes by the 1960s. Today they are still used in a few high power applications where their ability to withstand transient voltages and their robustness gives them an advantage over semiconductor devices. The recent (2012) resurgence of interest among audiophiles and recording studios in old valve audio gear such as guitar amplifiers and home audio systems has provided a market for the legacy consumer diode valves.

Semiconductor Diodes

Electronic Symbols

The symbol used for a semiconductor diode in a circuit diagram specifies the type of diode. There are alternative symbols for some types of diodes, though the differences are minor. The triangle in the symbols points to the forward direction, i.e. in the direction of current flow.

Diode

Light-emitting diode (LED)

Photodiode

Schottky diode

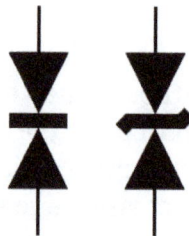

Transient-voltage-suppression diode (TVS)

Tunnel diode

Anode ⊳⊢⊢ Cathode

Varicap

Anode ▶⌐ Cathode

Zener diode

Anode (+) ⊳⊢ Cathode (−)

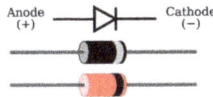

Typical diode packages in same alignment as diode symbol. Thin bar depicts the cathode.

A galena cat's-whisker detector, a point-contact diode.

Point-contact Diodes

A point-contact diode works the same as the junction diodes described below, but its construction is simpler. A pointed metal wire is placed in contact with an n-type semiconductor. Some metal migrates into the semiconductor to make a small p-type region around the contact. The 1N34 germanium version is still used in radio receivers as a detector and occasionally in specialized analog electronics.

Junction Diodes

p−n Junction Diode

A p−n junction diode is made of a crystal of semiconductor, usually silicon, but germanium and gallium arsenide are also used. Impurities are added to it to create a region on one side that contains negative charge carriers (electrons), called an n-type semiconductor, and a region on the other side that contains positive charge carriers (holes), called a p-type semiconductor. When the n-type and p-type materials are attached together, a momentary flow of electrons occur from the n to the p side resulting in a third region between the two where no charge carriers are present. This region is called the depletion region because there are no charge carriers (neither electrons nor holes) in it.

The diode's terminals are attached to the n-type and p-regions. The boundary between these two regions, called a p–n junction, is where the action of the diode takes place. When a sufficiently higher electrical potential is applied to the P side (the anode) than to the N side (the cathode), it allows electrons to flow through the depletion region from the N-type side to the P-type side. The junction does not allow the flow of electrons in the opposite direction when the potential is applied in reverse, creating, in a sense, an electrical check valve.

Schottky Diode

Another type of junction diode, the Schottky diode, is formed from a metal–semiconductor junction rather than a p–n junction, which reduces capacitance and increases switching speed.

Current–voltage Characteristic

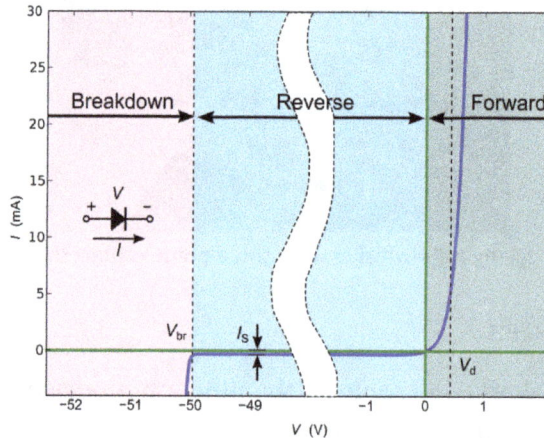

I–V (current vs. voltage) characteristics of a p–n junction diode

A semiconductor diode's behavior in a circuit is given by its current–voltage characteristic, or I–V graph. The shape of the curve is determined by the transport of charge carriers through the so-called *depletion layer* or *depletion region* that exists at the p–n junction between differing semiconductors. When a p–n junction is first created, conduction-band (mobile) electrons from the N-doped region diffuse into the P-doped region where there is a large population of holes (vacant places for electrons) with which the electrons "recombine". When a mobile electron recombines with a hole, both hole and electron vanish, leaving behind an immobile positively charged donor (dopant) on the N side and negatively charged acceptor (dopant) on the P side. The region around the p–n junction becomes depleted of charge carriers and thus behaves as an insulator.

However, the width of the depletion region (called the depletion width) cannot grow without limit. For each electron–hole pair recombination made, a positively charged dopant ion is left behind in the N-doped region, and a negatively charged dopant ion is created in the P-doped region. As recombination proceeds and more ions are created, an increasing electric field develops through the depletion zone that acts to slow and then finally stop recombination. At this point, there is a "built-in" potential across the depletion zone.

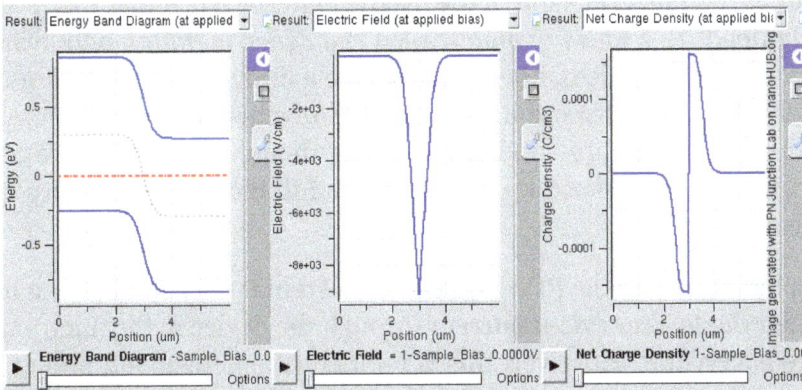

A PN junction diode in forward bias mode, the depletion width decreases. Both p and n junctions are doped at a 1e15/cm³ doping level, leading to built-in potential of ~0.59V. Observe the different Quasi Fermi levels for conduction band and valence band in n and p regions (red curves).

Reverse Bias

If an external voltage is placed across the diode with the same polarity as the built-in potential, the depletion zone continues to act as an insulator, preventing any significant electric current flow (unless electron–hole pairs are actively being created in the junction by, for instance, light). This is called the *reverse bias* phenomenon.

Forward Bias

However, if the polarity of the external voltage opposes the built-in potential, recombination can once again proceed, resulting in a substantial electric current through the p–n junction (i.e. substantial numbers of electrons and holes recombine at the junction). For silicon diodes, the built-in potential is approximately 0.7 V (0.3 V for germanium and 0.2 V for Schottky). Thus, if an external voltage greater than and opposite to the built-in voltage is applied, a current will flow and the diode is said to be "turned on" as it has been given an external *forward bias*. The diode is commonly said to have a forward "threshold" voltage, above which it conducts and below which conduction stops. However, this is only an approximation as the forward characteristic is according to the Shockley equation absolutely smooth.

A diode's I–V characteristic can be approximated by four regions of operation:

1. At very large reverse bias, beyond the peak inverse voltage or PIV, a process called reverse breakdown occurs that causes a large increase in current (i.e.,

a large number of electrons and holes are created at, and move away from the p–n junction) that usually damages the device permanently. The avalanche diode is deliberately designed for use in that manner. In the Zener diode, the concept of PIV is not applicable. A Zener diode contains a heavily doped p–n junction allowing electrons to tunnel from the valence band of the p-type material to the conduction band of the n-type material, such that the reverse voltage is "clamped" to a known value (called the *Zener voltage*), and avalanche does not occur. Both devices, however, do have a limit to the maximum current and power they can withstand in the clamped reverse-voltage region. Also, following the end of forward conduction in any diode, there is reverse current for a short time. The device does not attain its full blocking capability until the reverse current ceases.

2. For a bias less than the PIV, the reverse current is very small. For a normal P–N rectifier diode, the reverse current through the device in the micro-ampere (µA) range is very low. However, this is temperature dependent, and at sufficiently high temperatures, a substantial amount of reverse current can be observed (mA or more).

3. With a small forward bias, where only a small forward current is conducted, the current–voltage curve is exponential in accordance with the ideal diode equation. There is a definite forward voltage at which the diode starts to conduct significantly. This is called the *knee voltage* or *cut-in voltage* and is equal to the barrier potential of the p-n junction. This is a feature of the exponential curve, and appears sharper on a current scale more compressed than in the diagram shown here.

4. At larger forward currents the current-voltage curve starts to be dominated by the ohmic resistance of the bulk semiconductor. The curve is no longer exponential, it is asymptotic to a straight line whose slope is the bulk resistance. This region is particularly important for power diodes. The diode can be modeled as an ideal diode in series with a fixed resistor.

In a small silicon diode operating at its rated currents, the voltage drop is about 0.6 to 0.7 volts. The value is different for other diode types—Schottky diodes can be rated as low as 0.2 V, germanium diodes 0.25 to 0.3 V, and red or blue light-emitting diodes (LEDs) can have values of 1.4 V and 4.0 V respectively.

At higher currents the forward voltage drop of the diode increases. A drop of 1 V to 1.5 V is typical at full rated current for power diodes.

Shockley Diode Equation

The *Shockley ideal diode equation* or the *diode law* (named after the bipolar junction transistor co-inventor William Bradford Shockley) gives the I–V characteristic of an ideal diode in either forward or reverse bias (or no bias). The following equation is called the *Shockley ideal diode equation* when n, the ideality factor, is set equal to 1 :

$$I = I_S \left(e^{\frac{V_D}{nV_T}} - 1 \right)$$

where

I is the diode current,

I_S is the reverse bias saturation current (or scale current),

V_D is the voltage across the diode,

V_T is the thermal voltage, and

n is the *ideality factor*, also known as the *quality factor* or sometimes *emission coefficient*. The ideality factor n typically varies from 1 to 2 (though can in some cases be higher), depending on the fabrication process and semiconductor material and is set equal to 1 for the case of an "ideal" diode (thus the n is sometimes omitted). The ideality factor was added to account for imperfect junctions as observed in real transistors. The factor mainly accounts for carrier recombination as the charge carriers cross the depletion region.

The thermal voltage V_T is approximately 25.85 mV at 300 K, a temperature close to "room temperature" commonly used in device simulation software. At any temperature it is a known constant defined by:

$$V_T = \frac{kT}{q},$$

where k is the Boltzmann constant, T is the absolute temperature of the p–n junction, and q is the magnitude of charge of an electron (the elementary charge).

The reverse saturation current, I_S, is not constant for a given device, but varies with temperature; usually more significantly than V_T, so that V_D typically decreases as T increases.

The *Shockley ideal diode equation* or the *diode law* is derived with the assumption that the only processes giving rise to the current in the diode are drift (due to electrical field), diffusion, and thermal recombination–generation (R–G) (this equation is derived by setting n = 1 above). It also assumes that the R–G current in the depletion region is insignificant. This means that the *Shockley ideal diode equation* doesn't account for the processes involved in reverse breakdown and photon-assisted R–G. Additionally, it doesn't describe the "leveling off" of the I–V curve at high forward bias due to internal resistance. Introducing the ideality factor, n, accounts for recombination and generation of carriers.

Under *reverse bias* voltages the exponential in the diode equation is negligible, and the current is a constant (negative) reverse current value of $-I_S$. The reverse *breakdown region* is not modeled by the Shockley diode equation.

For even rather small *forward bias* voltages the exponential is very large, since the thermal voltage is very small in comparison. The subtracted '1' in the diode equation is then negligible and the forward diode current can be approximated by

$$I = I_S e^{\frac{V_D}{nV_T}}$$

Small-signal Behavior

For circuit design, a small-signal model of the diode behavior often proves useful. A specific example of diode modeling is discussed.

Reverse-Recovery Effect

Following the end of forward conduction in a p–n type diode, a reverse current can flow for a short time. The device does not attain its blocking capability until the mobile charge in the junction is depleted.

The effect can be significant when switching large currents very quickly. A certain amount of "reverse recovery time" t_r (on the order of tens of nanoseconds to a few microseconds) may be required to remove the reverse recovery charge Q_r from the diode. During this recovery time, the diode can actually conduct in the reverse direction. This might give rise to a large constant current in the reverse direction for a short time while the diode is reverse biased. The magnitude of such a reverse current is determined by the operating circuit (i.e., the series resistance) and the diode is said to be in the storage-phase. In certain real-world cases it is important to consider the losses that are incurred by this non-ideal diode effect. However, when the slew rate of the current is not so severe (e.g. Line frequency) the effect can be safely ignored. For most applications, the effect is also negligible for Schottky diodes.

The reverse current ceases abruptly when the stored charge is depleted; this abrupt stop is exploited in step recovery diodes for generation of extremely short pulses.

Types of Semiconductor Diode

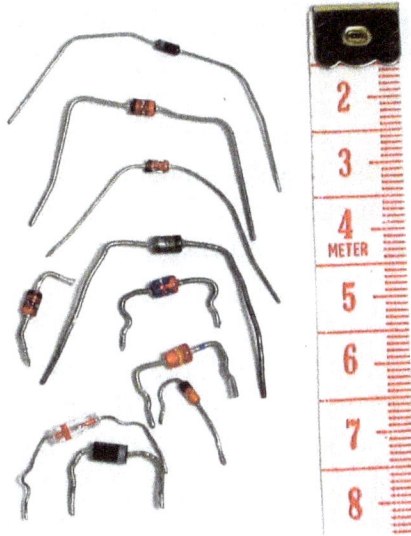

Several types of diodes. The scale is centimeters.

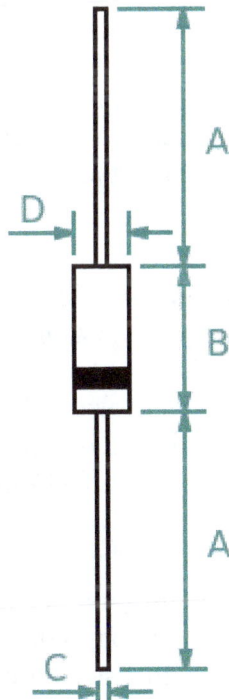

Typical datasheet drawing showing the dimensions of a DO-41 diode package

There are several types of p–n junction diodes, which emphasize either a different physical aspect of a diode often by geometric scaling, doping level, choosing the right electrodes, are just an application of a diode in a special circuit, or are really different devices like the Gunn and laser diode and the MOSFET:

Normal (p–n) diodes, which operate as described above, are usually made of doped silicon or, more rarely, germanium. Before the development of silicon power rectifier diodes, cuprous oxide and later selenium was used. Their low efficiency required a much higher forward voltage to be applied (typically 1.4 to 1.7 V per "cell", with multiple cells stacked so as to increase the peak inverse voltage rating for application in high voltage rectifiers), and required a large heat sink (often an extension of the diode's metal substrate), much larger than the later silicon diode of the same current ratings would require. The vast majority of all diodes are the p–n diodes found in CMOS integrated circuits, which include two diodes per pin and many other internal diodes.

Avalanche diodes

> These are diodes that conduct in the reverse direction when the reverse bias voltage exceeds the breakdown voltage. These are electrically very similar to Zener diodes (and are often mistakenly called Zener diodes), but break down by a different mechanism: the *avalanche effect*. This occurs when the reverse electric field applied across the p–n junction causes a wave of ionization, reminiscent of an avalanche, leading to a large current. Avalanche diodes are designed to break down at a well-defined reverse voltage without being destroyed. The difference between the avalanche diode (which has a reverse breakdown above about 6.2 V) and the Zener is that the channel length of the former exceeds the mean free path of the electrons, resulting in many collisions between them on the way through the channel. The only practical difference between the two types is they have temperature coefficients of opposite polarities.

Cat's whisker or crystal diodes

> These are a type of point-contact diode. The cat's whisker diode consists of a thin or sharpened metal wire pressed against a semiconducting crystal, typically galena or a piece of coal. The wire forms the anode and the crystal forms the cathode. Cat's whisker diodes were also called crystal diodes and found application in the earliest radios called crystal radio receivers. Cat's whisker diodes are generally obsolete, but may be available from a few manufacturers.

Constant current diodes

> These are actually JFETs with the gate shorted to the source, and function like a two-terminal current-limiting analog to the voltage-limiting Zener diode. They allow a current through them to rise to a certain value, and then level off at a specific value. Also called *CLDs*, *constant-current diodes*, *diode-connected transistors*, or *current-regulating diodes*.

Esaki or tunnel diodes

> These have a region of operation showing negative resistance caused by quantum tunneling, allowing amplification of signals and very simple bistable circuits.

Because of the high carrier concentration, tunnel diodes are very fast, may be used at low (mK) temperatures, high magnetic fields, and in high radiation environments. Because of these properties, they are often used in spacecraft.

Gunn diodes

These are similar to tunnel diodes in that they are made of materials such as GaAs or InP that exhibit a region of negative differential resistance. With appropriate biasing, dipole domains form and travel across the diode, allowing high frequency microwave oscillators to be built.

Light-emitting diodes (LEDs)

In a diode formed from a direct band-gap semiconductor, such as gallium arsenide, charge carriers that cross the junction emit photons when they recombine with the majority carrier on the other side. Depending on the material, wavelengths (or colors) from the infrared to the near ultraviolet may be produced. The forward potential of these diodes depends on the wavelength of the emitted photons: 2.1 V corresponds to red, 4.0 V to violet. The first LEDs were red and yellow, and higher-frequency diodes have been developed over time. All LEDs produce incoherent, narrow-spectrum light; "white" LEDs are actually combinations of three LEDs of a different color, or a blue LED with a yellow scintillator coating. LEDs can also be used as low-efficiency photodiodes in signal applications. An LED may be paired with a photodiode or phototransistor in the same package, to form an opto-isolator.

Laser diodes

When an LED-like structure is contained in a resonant cavity formed by polishing the parallel end faces, a laser can be formed. Laser diodes are commonly used in optical storage devices and for high speed optical communication.

Thermal diodes

This term is used both for conventional p–n diodes used to monitor temperature because of their varying forward voltage with temperature, and for Peltier heat pumps for thermoelectric heating and cooling. Peltier heat pumps may be made from semiconductor, though they do not have any rectifying junctions, they use the differing behaviour of charge carriers in N and P type semiconductor to move heat.

Perun's diodes

This is a special type of voltage-surge protection diode. It is characterized by the symmetrical voltage-current characteristic, similar to DIAC. It has much faster response time however, that's why it is used in demanding applications.

Photodiodes

All semiconductors are subject to optical charge carrier generation. This is typically an undesired effect, so most semiconductors are packaged in light blocking material. Photodiodes are intended to sense light(photodetector), so they are packaged in materials that allow light to pass, and are usually PIN (the kind of diode most sensitive to light). A photodiode can be used in solar cells, in photometry, or in optical communications. Multiple photodiodes may be packaged in a single device, either as a linear array or as a two-dimensional array. These arrays should not be confused with charge-coupled devices.

PIN diodes

A PIN diode has a central un-doped, or *intrinsic*, layer, forming a p-type/intrinsic/n-type structure. They are used as radio frequency switches and attenuators. They are also used as large-volume, ionizing-radiation detectors and as photodetectors. PIN diodes are also used in power electronics, as their central layer can withstand high voltages. Furthermore, the PIN structure can be found in many power semiconductor devices, such as IGBTs, power MOSFETs, and thyristors.

Schottky diodes

Schottky diodes are constructed from a metal to semiconductor contact. They have a lower forward voltage drop than p–n junction diodes. Their forward voltage drop at forward currents of about 1 mA is in the range 0.15 V to 0.45 V, which makes them useful in voltage clamping applications and prevention of transistor saturation. They can also be used as low loss rectifiers, although their reverse leakage current is in general higher than that of other diodes. Schottky diodes are majority carrier devices and so do not suffer from minority carrier storage problems that slow down many other diodes—so they have a faster reverse recovery than p–n junction diodes. They also tend to have much lower junction capacitance than p–n diodes, which provides for high switching speeds and their use in high-speed circuitry and RF devices such as switched-mode power supply, mixers, and detectors.

Super barrier diodes

Super barrier diodes are rectifier diodes that incorporate the low forward voltage drop of the Schottky diode with the surge-handling capability and low reverse leakage current of a normal p–n junction diode.

Gold-doped diodes

As a dopant, gold (or platinum) acts as recombination centers, which helps a fast recombination of minority carriers. This allows the diode to operate at signal frequencies, at the expense of a higher forward voltage drop. Gold-doped diodes are faster than other p–n diodes (but not as fast as Schottky diodes). They also have less reverse-current leakage than Schottky diodes (but not as good as other p–n diodes). A typical example is the 1N914.

Snap-off or Step recovery diodes

The term *step recovery* relates to the form of the reverse recovery characteristic of these devices. After a forward current has been passing in an SRD and the current is interrupted or reversed, the reverse conduction will cease very abruptly (as in a step waveform). SRDs can, therefore, provide very fast voltage transitions by the very sudden disappearance of the charge carriers.

Stabistors or *Forward Reference Diodes*

The term *stabistor* refers to a special type of diodes featuring extremely stable forward voltage characteristics. These devices are specially designed for low-voltage stabilization applications requiring a guaranteed voltage over a wide current range and highly stable over temperature.

Transient voltage suppression diode (TVS)

These are avalanche diodes designed specifically to protect other semiconductor devices from high-voltage transients. Their p–n junctions have a much larger cross-sectional area than those of a normal diode, allowing them to conduct large currents to ground without sustaining damage.

Varicap or varactor diodes

These are used as voltage-controlled capacitors. These are important in PLL (phase-locked loop) and FLL (frequency-locked loop) circuits, allowing tuning circuits, such as those in television receivers, to lock quickly on to the frequency. They also enabled tunable oscillators in early discrete tuning of radios, where a cheap and stable, but fixed-frequency, crystal oscillator provided the reference frequency for a voltage-controlled oscillator.

Zener diodes

These can be made to conduct in reverse bias (backward), and are correctly termed reverse breakdown diodes. This effect, called Zener breakdown, occurs at a precisely defined voltage, allowing the diode to be used as a precision voltage reference. The term Zener diode is colloquially applied to several types of breakdown diodes, but strictly speaking Zener diodes have a breakdown voltage of below 5 volts, whilst avalanche diodes are used for breakdown voltages above that value. In practical voltage reference circuits, Zener and switching diodes are connected in series and opposite directions to balance the temperature coefficient response of the diodes to near-zero. Some devices labeled as high-voltage Zener diodes are actually avalanche diodes. Two (equivalent) Zeners in series and in reverse order, in the same package, constitute a transient absorber (or Transorb, a registered trademark).

Other uses for semiconductor diodes include the sensing of temperature, and computing analog logarithms.

Numbering and Coding Schemes

There are a number of common, standard and manufacturer-driven numbering and coding schemes for diodes; the two most common being the EIA/JEDEC standard and the European Pro Electron standard:

EIA/JEDEC

The standardized 1N-series numbering *EIA370* system was introduced in the US by EIA/JEDEC (Joint Electron Device Engineering Council) about 1960. Most diodes have a 1-prefix designation (e.g., 1N4003). Among the most popular in this series were: 1N34A/1N270 (germanium signal), 1N914/1N4148 (silicon signal), 1N400x (silicon 1A power rectifier), and 1N580x (silicon 3A power rectifier).

JIS

The JIS semiconductor designation system has all semiconductor diode designations starting with "1S".

Pro Electron

The European Pro Electron coding system for active components was introduced in 1966 and comprises two letters followed by the part code. The first letter represents the semiconductor material used for the component (A = germanium and B = silicon) and the second letter represents the general function of the part (for diodes, A = low-power/signal, B = variable capacitance, X = multiplier, Y = rectifier and Z = voltage reference); for example:

- AA-series germanium low-power/signal diodes (e.g., AA119)
- BA-series silicon low-power/signal diodes (e.g., BAT18 silicon RF switching diode)
- BY-series silicon rectifier diodes (e.g., BY127 1250V, 1A rectifier diode)
- BZ-series silicon Zener diodes (e.g., BZY88C4V7 4.7V Zener diode)

Other common numbering / coding systems (generally manufacturer-driven) include:

- GD-series germanium diodes (e.g., GD9) – this is a very old coding system

- OA-series germanium diodes (e.g., OA47) – a coding sequence developed by Mullard, a UK company

As well as these common codes, many manufacturers or organisations have their own systems too – for example:

- HP diode 1901-0044 = JEDEC 1N4148

- UK military diode CV448 = Mullard type OA81 = GEC type GEX23

Related Devices

- Rectifier

- Transistor

- Thyristor or silicon controlled rectifier (SCR)

- TRIAC

- DIAC

- Varistor

In optics, an equivalent device for the diode but with laser light would be the Optical isolator, also known as an Optical Diode, that allows light to only pass in one direction. It uses a Faraday rotator as the main component.

Applications

Radio Demodulation

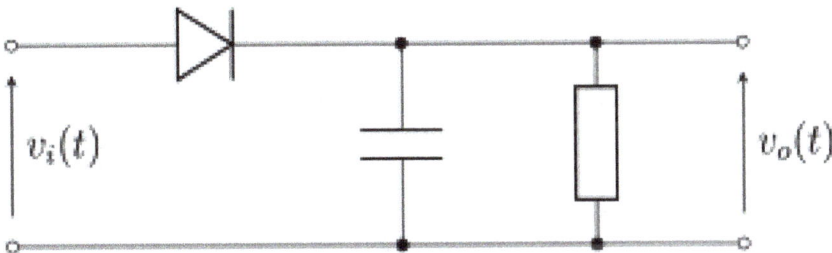

A simple envelope demodulator circuit.

The first use for the diode was the demodulation of amplitude modulated (AM) radio broadcasts. An AM signal consists of alternating positive and negative peaks of a radio carrier wave, whose amplitude or envelope is proportional to the original audio signal. The diode (originally a crystal diode) rectifies the AM radio frequency signal, leaving only the positive peaks of the carrier wave. The audio is then extracted from the rectified carrier wave using a simple filter and fed into an audio amplifier or transducer, which generates sound waves.

Power Conversion

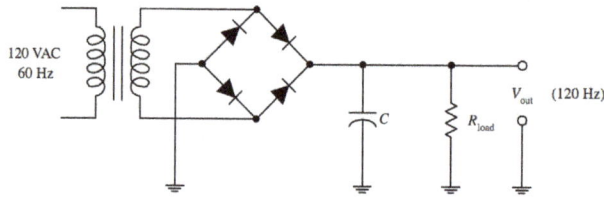

120 VAC
60 Hz

V_{out} (120 Hz)

C

R_{load}

Schematic of basic AC-to-DC power supply

Rectifiers are constructed from diodes, where they are used to convert alternating current (AC) electricity into direct current (DC). Automotive alternators are a common example, where the diode, which rectifies the AC into DC, provides better performance than the commutator or earlier, dynamo. Similarly, diodes are also used in *Cockcroft–Walton voltage multipliers* to convert AC into higher DC voltages.

Over-voltage Protection

Diodes are frequently used to conduct damaging high voltages away from sensitive electronic devices. They are usually reverse-biased (non-conducting) under normal circumstances. When the voltage rises above the normal range, the diodes become forward-biased (conducting). For example, diodes are used in (stepper motor and H-bridge) motor controller and relay circuits to de-energize coils rapidly without the damaging voltage spikes that would otherwise occur. (A diode used in such an application is called a flyback diode). Many integrated circuits also incorporate diodes on the connection pins to prevent external voltages from damaging their sensitive transistors. Specialized diodes are used to protect from over-voltages at higher power.

Logic Gates

Diodes can be combined with other components to construct AND and OR logic gates. This is referred to as diode logic.

Ionizing Radiation Detectors

In addition to light, mentioned above, semiconductor diodes are sensitive to more energetic radiation. In electronics, cosmic rays and other sources of ionizing radiation cause noise pulses and single and multiple bit errors. This effect is sometimes exploited by particle detectors to detect radiation. A single particle of radiation, with thousands or millions of electron volts of energy, generates many charge carrier pairs, as its energy is deposited in the semiconductor material. If the depletion layer is large enough to catch the whole shower or to stop a heavy particle, a fairly accurate measurement of the particle's energy can be made, simply by measuring the charge conducted and without the complexity of a magnetic spectrometer, etc.

These semiconductor radiation detectors need efficient and uniform charge collection and low leakage current. They are often cooled by liquid nitrogen. For longer-range (about a centimetre) particles, they need a very large depletion depth and large area. For short-range particles, they need any contact or un-depleted semiconductor on at least one surface to be very thin. The back-bias voltages are near breakdown (around a thousand volts per centimetre). Germanium and silicon are common materials. Some of these detectors sense position as well as energy. They have a finite life, especially when detecting heavy particles, because of radiation damage. Silicon and germanium are quite different in their ability to convert gamma rays to electron showers.

Semiconductor detectors for high-energy particles are used in large numbers. Because of energy loss fluctuations, accurate measurement of the energy deposited is of less use.

Temperature Measurements

A diode can be used as a temperature measuring device, since the forward voltage drop across the diode depends on temperature, as in a silicon bandgap temperature sensor. From the Shockley ideal diode equation given above, it might *appear* that the voltage has a *positive* temperature coefficient (at a constant current), but usually the variation of the reverse saturation current term is more significant than the variation in the thermal voltage term. Most diodes therefore have a *negative* temperature coefficient, typically –2 mV/°C for silicon diodes. The temperature coefficient is approximately constant for temperatures above about 20 kelvins. Some graphs are given for 1N400x series, and CY7 cryogenic temperature sensor.

Current Steering

Diodes will prevent currents in unintended directions. To supply power to an electrical circuit during a power failure, the circuit can draw current from a battery. An uninterruptible power supply may use diodes in this way to ensure that current is only drawn from the battery when necessary. Likewise, small boats typically have two circuits each with their own battery/batteries: one used for engine starting; one used for domestics. Normally, both are charged from a single alternator, and a heavy-duty split-charge diode is used to prevent the higher-charge battery (typically the engine battery) from discharging through the lower-charge battery when the alternator is not running.

Diodes are also used in electronic musical keyboards. To reduce the amount of wiring needed in electronic musical keyboards, these instruments often use keyboard matrix circuits. The keyboard controller scans the rows and columns to determine which note the player has pressed. The problem with matrix circuits is that, when several notes are pressed at once, the current can flow backwards through the circuit and trigger "phantom keys" that cause "ghost" notes to play. To avoid triggering unwanted notes, most keyboard matrix circuits have diodes soldered with the switch under each key of the musical keyboard. The same principle is also used for the switch matrix in solid-state pinball machines.

Waveform Clipper

Diodes can be used to limit the positive or negative excursion of a signal to a prescribed voltage.

Clamper

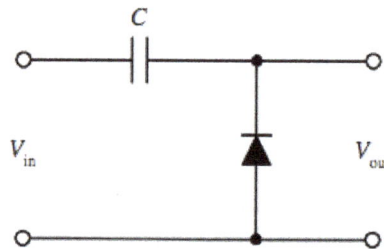

This simple diode clamp will clamp the negative peaks of the incoming waveform to the common rail voltage

A diode clamp circuit can take a periodic alternating current signal that oscillates between positive and negative values, and vertically displace it such that either the positive, or the negative peaks occur at a prescribed level. The clamper does not restrict the peak-to-peak excursion of the signal, it moves the whole signal up or down so as to place the peaks at the reference level.

Abbreviations

Diodes are usually referred to as D for diode on PCBs. Sometimes the abbreviation CR for *crystal rectifier* is used.

Backward Diode

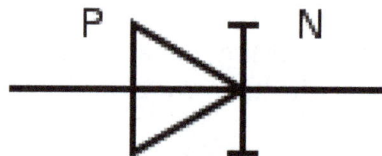

The schematic symbol for the backward diode, annotated to show which side is P type and which is N; current flows most easily from N to P, backward relative to the arrow.

In semiconductor devices, a backward diode (also called back diode) is a variation on a Zener diode or tunnel diode having a better conduction for small reverse biases (for example −0.1 to −0.6 V) than for forward bias voltages.

The reverse current in such a diode is by tunneling, which is also known as the *tunnel effect*.

Current–voltage Characteristics of Backward Diode

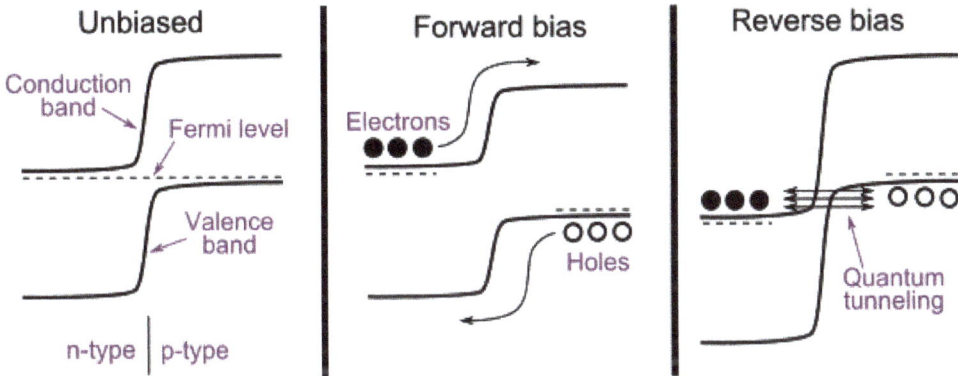

Band diagram of a backward diode. Electron energy is on the vertical axis, position within the device is on the horizontal axis. The backward diode has the unusual property that the so-called reverse bias direction actually has more current flow than the so-called forward bias.

The forward I–V characteristic is the same as that of an ordinary P–N diode. The breakdown starts when reverse voltage is applied. In the case of Zener breakdown, it starts at a particular voltage. In this diode the voltage remains relatively constant (independent of current) when it is connected in reverse bias. The backward diode is a special form of tunnel diode in which the tunneling phenomenon is only incipient, and the negative resistance region virtually disappears. The forward current is very small and becomes equivalent to the reverse current of a conventional diode.

Applications of Backward Diodes

Detector

Since it has low capacitance and no charge storage effect, and a strongly nonlinear small-signal characteristic, the backward diode can be used as a detector up to 40 GHz.

Rectifier

A backward diode can be used for rectifying weak signals with peak amplitudes of 0.1 to 0.7 V.

Switch

A backward diode can be used in high speed switching applications.

Space Charge Capacitance C_T of Diode

Reverse bias causes majority carriers to move away from the junction, thereby creating more ions. Hence the thickness of depletion region increases. This region behaves as the dielectric material used for making capacitors. The p-type and n-type conducting on each side of dielectric act as the plate. The incremental capacitance C_T is defined by

$$C_T = \left| \frac{dQ}{dV} \right|$$

Since $\quad i = \dfrac{dQ}{dt}$

Therefore, $\mathrm{i} = C_T \dfrac{dV}{dt}$ $\qquad$ (E-1)

where, dQ is the increase in charge caused by a change dV in voltage. C_T is not constant, it depends upon applied voltage, there fore it is defined as dQ / dV.

When p-n junction is forward biased, then also a capacitance is defined called *diffusion capacitance* C_D (rate of change of injected charge with voltage) to take into account the time delay in moving the charges across the junction by the diffusion process. It is considered as a fictitious element that allow us to predict time delay.

If the amount of charge to be moved across the junction is increased, the time delay is greater, it follows that diffusion capacitance varies directly with the magnitude of forward current.

$$C_D = \frac{dQ}{dV} = \frac{I\tau}{dV} \qquad \text{(E-2)}$$

Relationship Between Diode Current and Diode Voltage

An exponential relationship exists between the carrier density and applied potential of diode junction as given in equation E-3. This exponential relationship of the current i_D and the voltage v_D holds over a range of at least seven orders of magnitudes of current - that is a factor of 10^7.

$$i_D = I_0 \left[\exp\left(\frac{qv_D}{nkT} \right) - 1 \right] - I_0 \left[e^{\left(\frac{qv_D}{nkT} \right)} - 1 \right] \qquad \text{(E-3)}$$

Where,

i_D = Current through the diode (dependent variable in this expression)

v_D = Potential difference across the diode terminals (independent variable in this expression)

I_0 = Reverse saturation current (of the order of 10^{-15} A for small signal diodes, but I_0 is a strong function of temperature)

q = Electron charge: 1.60×10^{-19} joules/volt

k = Boltzmann's constant: 1.38×10^{-23} joules /° K

T = Absolute temperature in degrees Kelvin (°K = 273 + temperature in °C)

n = Empirical scaling constant between 0.5 and 2, sometimes referred to as the Exponential Ideality Factor

The empirical constant, n, is a number that can vary according to the voltage and current levels. It depends on electron drift, diffusion, and carrier recombination in the depletion region. Among the quantities affecting the value of n are the diode manufacture, levels of doping and purity of materials. If $n=1$, the value of $k\,T/q$ is 26 mV at 25°C. When $n=2$, $k\,T/q$ becomes 52 mV.

For germanium diodes, n is usually considered to be close to 1. For silicon diodes, n is in the range of 1.3 to 1.6. n is assumed 1 for all junctions all throughout unless otherwise noted.

Equation (E-3) can be simplified by defining $V_T = k\,T/q$, yielding

$$i_D = I_0\left[\exp\left(\frac{v_D}{nVT}\right) - 1\right] - I_0\left[e^{\left(\frac{v_D}{nVT}\right)} - 1\right] \qquad \text{(E-4)}$$

At room temperature (25°C) with forward-bias voltage only the first term in the parentheses is dominant and the current is approximately given by

$$i_D = I_0\, e^{\frac{v_D}{nV_T}} \qquad \text{(E-5)}$$

The current-voltage (l-V) characteristic of the diode, as defined by (E-3) is illustrated in the figure given below. The curve in the figure consists of two exponential curves. However, the exponent values are such that for voltages and currents experienced in practical circuits, the curve sections are close to being straight lines. For voltages less than V_{ON}, the curve is approximated by a straight line of slope close to zero. Since the slope is the conductance (i.e., i/v), the conductance is very small in this region, and the equivalent resistance is very high. For voltages above V_{ON}, the curve is approximated by a straight line with a very large slope. The conductance is therefore very large, and the diode has a very small equivalent resistance.

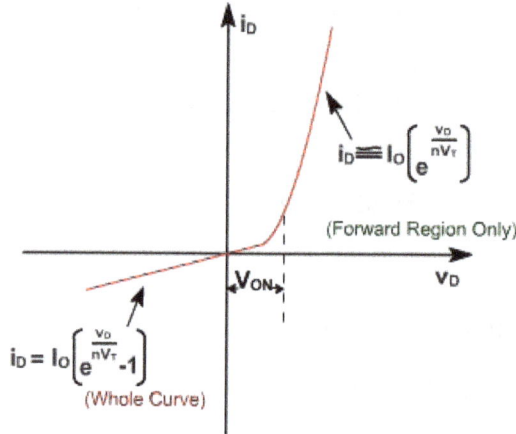

Diode Voltage relationship

The slope of the curves changes as the current and voltage change since the l-V characteristic follows the exponential relationship of relationship of equation (E-4). Differentiate the equation (E-4) to find the slope at any arbitrary value of v_D or i_D,

$$\frac{di_D}{dv_D} = \frac{I_0}{nV_T}\exp\left(\frac{v_D}{nV_T}\right) = \frac{I_0}{nV_T}e^{\frac{v_D}{nV_r}} \qquad \text{(E-6)}$$

This slope is the equivalent conductance of the diode at the specified values of v_D or i_D.

We can approximate the slope as a linear function of the diode current. To eliminate the exponential function, we substitute equation (E-4) into the exponential of equation (E-7) to obtain

$$\exp\left(\frac{v_D}{nV_T}\right) = \frac{i_D}{I_0} + 1 = \left(\frac{di_D}{dv_D}\right)\left(\frac{nV_T}{i_0}\right) \qquad \text{(E-7)}$$

A realistic assumption is that $I_{0<<}i_D$ equation (E-7) then yields,

$$\frac{di_D}{dv_D} = \frac{i_D + I_0}{nV_T} \approx \frac{i_D}{nV_T} \qquad \text{(E-8)}$$

The approximation applies if the diode is forward biased. The dynamic resistance is the reciprocal of this expression.

$$r_d = \frac{nV_T}{i_D + I_0} \approx \frac{nV_T}{i_D} \qquad \text{(E-9)}$$

Although r_d is a function of i_d, we can approximate it as a constant if the variation of i_D is small. This corresponds to approximating the exponential function as a straight line within a specific operating range.

Normally, the term R_f to denote diode forward resistance. R_f is composed of r_d and the contact resistance. The contact resistance is a relatively small resistance composed of the resistance of the actual connection to the diode and the resistance of the semiconductor prior to the junction. The reverse-bias resistance is extremely large and is often approximated as infinity.

Temperature Effects

Temperature plays an important role in determining the characteristic of diodes. As temperature increases, the turn-on voltage, v_{ON}, decreases. Alternatively, a decrease in temperature results in an increase in v_{ON}. This is illustrated in the figure, where V_{ON} varies linearly with temperature which is evidenced by the evenly spaced curves for increasing temperature in 25 °C increments.

The temperature relationship is described by equation

$$V_{ON}(T_{New}) - V_{ON}(T_{room}) = k_T(T_{New} - T_{room}) \qquad \text{(E-10)}$$

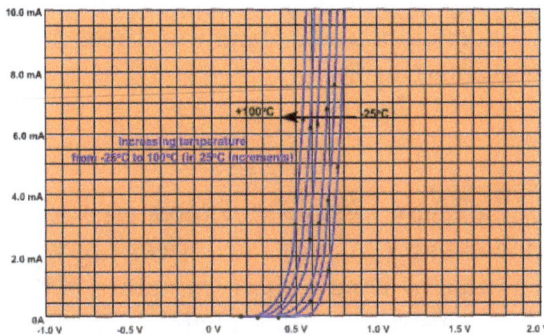

Dependence of iD on temperature versus vD for real diode (kT = -2.0 mV /°C)

where,

   T_{room} = room temperature, or 25°C.

   T_{New} = new temperature of diode in °C.

$V_{ON}(T_{room})$ = diode voltage at room temperature.

$V_{ON}(T_{New})$ = diode voltage at new temperature.

   k_T = temperature coefficient in V/°C.

Although k_T varies with changing operating parameters, standard engineering practice permits approximation as a constant. Values of k_T for the various types of diodes at room temperature are given as follows:

k_T = -2.5 mV/°C for germanium diodes

k_T = -2.0 mV/°C for silicon diodes

The reverse saturation current, I_0 also depends on temperature. At room temperature, it increases approximately 16% per °C for silicon and 10% per °C for germanium diodes. In other words, I_0 approximately doubles for every 5 °C increase in temperature for silicon, and for every 7 °C for germanium. The expression for the reverse saturation current as a function of temperature can be approximated as

$$I_0(atT_2) = I_0(atT_1)\exp(k_i(T_2 - T_1)) = I_0(atT_1)e^{K_i(T_2 - T_1)} \quad \text{(E-11)}$$

where $K_i = 0.15/°C$ (for silicon) and T1 and T2 are two arbitrary temperatures.

p–n Diode

A p–n diode is a type of semiconductor diode based upon the p–n junction. The diode conducts current in only one direction, and it is made by joining a *p*-type semiconducting layer to an *n*-type semiconducting layer. Semiconductor diodes have multiple uses including rectification of alternating current to direct current, detection of radio signals, emitting light and detecting light.

Structure

The figure shows two of the many possible structures used for *p–n* semiconductor diodes, both adapted to increase the voltage the devices can withstand in reverse bias. The top structure uses a mesa to avoid a sharp curvature of the *p+*-region next to the adjoining *n*-layer. The bottom structure uses a lightly doped *p*-guard-ring at the edge of the sharp corner of the *p+*-layer to spread the voltage out over a larger distance and reduce the electric field. (Superscripts like n^+ or n^- refer to heavier or lighter impurity doping levels.)

Electrical Behavior

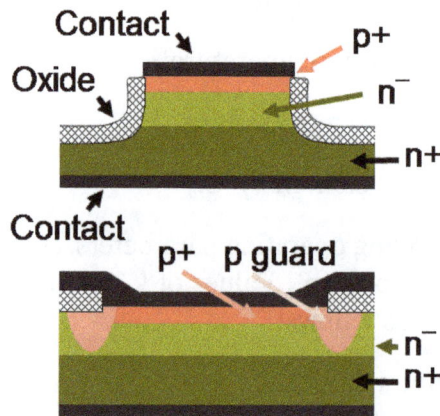

Mesa diode structure (top) and planar diode structure with guard-ring (bottom).

The ideal diode has zero resistance for the *forward bias polarity*, and infinite resistance (conducts zero current) for the *reverse voltage polarity*; if connected in an alternating current circuit, the semiconductor diode acts as an *electrical rectifier*.

The semiconductor diode is not ideal. As shown in the figure, the diode does not conduct appreciably until a nonzero *knee voltage* (also called the *turn-on voltage* or the *cut-in voltage*) is reached. Above this voltage the slope of the current-voltage curve is not infinite (on-resistance is not zero). In the reverse direction the diode conducts a nonzero leakage current (exaggerated by a smaller scale in the figure) and at a sufficiently large reverse voltage below the *breakdown voltage* the current increases very rapidly with more negative reverse voltages.

As shown in the figure, the *on* and *off* resistances are the reciprocal slopes of the current-voltage characteristic at a selected bias point:

$$r_D = \frac{\Delta v_D}{\Delta i_D}\bigg|_{v_D=V_{BIAS}} ,$$

where r_D is the resistance and Δi_D is the current change corresponding to the diode voltage change Δv_D at the bias $v_D=V_{BIAS}$.

Operation

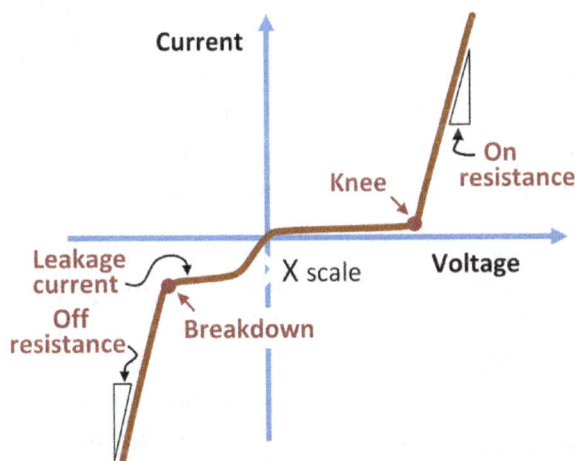

Nonideal *p–n* diode current-voltage characteristics.

Here, the operation of the abrupt *p–n* diode is considered. By "abrupt" is meant that the p- and n-type doping exhibit a step function discontinuity at the plane where they encounter each other. The objective is to explain the various bias regimes in the figure displaying current-voltage characteristics. Operation is described using band-bending diagrams that show how the lowest conduction band energy and the highest valence band energy vary with position inside the diode under various bias conditions.

Zero Bias

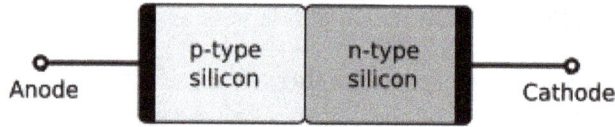

An abrupt p–n diode made by doping silicon.

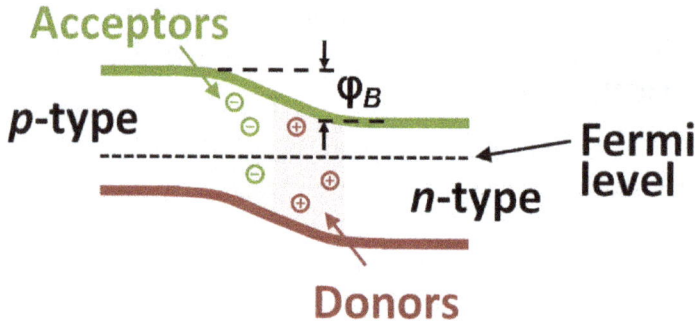

Band-bending diagram for p–n diode at zero applied voltage. The depletion region is shaded.

The figure shows a band bending diagram for a p–n diode; that is, the band edges for the conduction band (upper line) and the valence band (lower line) are shown as a function of position on both sides of the junction between the p-type material (left side) and the n-type material (right side). When a p-type and an n-type region of the same semiconductor are brought together and the two diode contacts are short-circuited, the Fermi half-occupancy level (dashed horizontal straight line) is situated at a constant level. This level ensures that in the field-free bulk on both sides of the junction the hole and electron occupancies are correct. (So, for example, it is not necessary for an electron to leave the n-side and travel to the p-side through the short circuit to adjust the occupancies.)

However, a flat Fermi level requires the bands on the p-type side to move higher than the corresponding bands on the n-type side, forming a step or barrier in the band edges, labeled φ_B. This step forces the electron density on the p-side to be a Boltzmann factor $\exp(-\varphi_B/V_{th})$ smaller than on the n-side, corresponding to the lower electron density in p-region. The symbol V_{th} denotes the *thermal voltage*, defined as $V_{th} = k_B T/q$. At $T = 290$ kelvins (room temperature), the thermal voltage is approximately 25 mV. Similarly, hole density on the n-side is a Boltzmann factor smaller than on the p-side. This reciprocal reduction in minority carrier density across the junction forces the pn-product of carrier densities to be

$$pn = p_B n_B e^{-\varphi_B/V_{th}}$$

at any position within the diode at equilibrium. Where p_B and n_B are the bulk majority carrier densities on the p-side and the n-side, respectively.

As a result of this step in band edges, a *depletion region* near the junction becomes depleted of both holes and electrons, forming an insulating region with almost no *mobile* charges. There are, however, *fixed, immobile* charges due to dopant ions. The near absence of mobile charge in the depletion layer means that the mobile charges present are insufficient to balance the immobile charge contributed by the dopant ions: a negative charge on the *p*-type side due to acceptor dopant and as a positive charge on the *n*-type side due to donor dopant. Because of this charge there is an electric field in this region, as determined by Poisson's equation. The width of the depletion region adjusts so the negative acceptor charge on the *p*-side exactly balances the positive donor charge on the *n*-side, so there is no electric field outside the depletion region on either side.

In this band configuration no voltage is applied and no current flows through the diode. To force current through the diode a *forward bias* must be applied, as described next.

Forward Bias

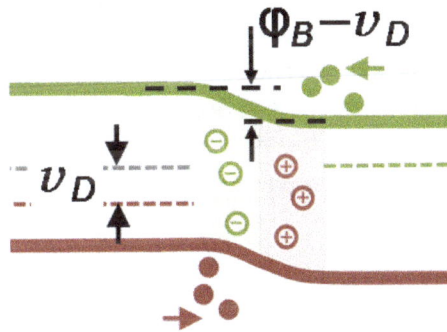

Band-bending diagram for *p–n* diode in forward bias. Diffusion drives carriers across the junction.

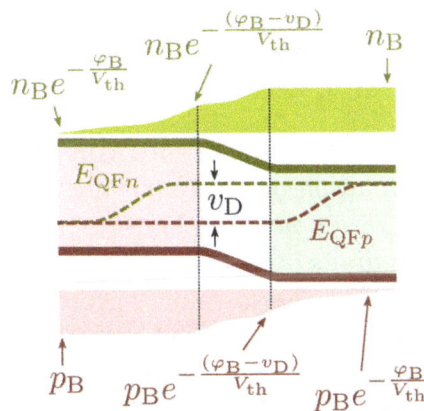

Quasi-Fermi levels and carrier densities in forward biased *p–n-* diode. The figure assumes recombination is confined to the regions where majority carrier concentration is near the bulk values, which is not accurate when recombination-generation centers in the field region play a role.

In forward bias, positive terminal of the battery is connected to the *p*- type material and negative terminal is connected to the *n*- type material so that holes are injected into the p-type material and electrons into the n-type material. The electrons in the n-type material are called *majority* carriers on that side, but electrons that make it to the p-type side are called *minority* carriers. The same descriptors apply to holes: they are majority carriers on the p-type side, and minority carriers on the n-type side.

A forward bias separates the two bulk half-occupancy levels by the amount of the applied voltage, which lowers the separation of the p-type bulk band edges to be closer in energy to those of the n-type. As shown in the diagram, the step in band edges is reduced by the applied voltage to $\varphi_B - v_D$. (The band bending diagram is made in units of volts, so no electron charge appears to convert v_D to energy.)

Under forward bias, a *diffusion current* flows (that is a current driven by a concentration gradient) of holes from the p-side into the n-side, and of electrons in the opposite direction from the n-side to the p-side. The gradient driving this transfer is set up as follows: in the bulk distant from the interface, minority carriers have a very low concentration compared to majority carriers, for example, electron density on the p-side (where they are minority carriers) is a factor $\exp(-\varphi_B/V_{th})$ lower than on the n-side (where they are majority carriers). On the other hand, near the interface, application of voltage v_D reduces the step in band edges and increases minority carrier densities by a Boltzmann factor $\exp(v_D/V_{th})$ above the bulk values. Within the junction, the pn-product is increased above the equilibrium value to:

$$pn = \left(p_B n_B \, e^{-\varphi_B/V_{th}} \right) e^{v_D/V_{th}} \, .$$

The gradient driving the diffusion is then the difference between the large excess minority carrier densities at the barrier and the low densities in the bulk, and that gradient drives diffusion of minority carriers from the interface into the bulk. The injected minority carriers are reduced in number as they travel into the bulk by *recombination* mechanisms that drive the excess concentrations toward the bulk values.

Recombination can occur by direct encounter with a majority carrier, annihilating both carriers, or through a *recombination-generation* center, a defect that alternately traps holes and electrons, assisting recombination. The minority carriers have a limited *lifetime*, and this lifetime in turn limits how far they can diffuse from the majority carrier side into the minority carrier side, the so-called *diffusion length*. In the LED recombination of electrons and holes is accompanied by emission of light of a wavelength related to the energy gap between valence and conduction bands, so the diode converts a portion of the forward current into light.

Under forward bias, the half-occupancy lines for holes and electrons cannot remain flat throughout the device as they are when in equilibrium, but become *quasi-Fermi levels* that vary with position. As shown in the figure, the electron quasi-Fermi level shifts with position, from the half-occupancy equilibrium Fermi level in the n-bulk, to the half-occupancy equilibrium level for holes deep in the p-bulk. The hole quasi-Fermi level does the reverse. The two quasi-Fermi levels do not coincide except deep in the bulk materials.

The figure shows the majority carrier densities drop from the majority carrier density levels n_B, p_B in their respective bulk materials, to a level a factor $\exp(-(\varphi_B-v_D)/V_{th})$ smaller at the top of the barrier, which is reduced from the equilibrium value φ_B by the amount of the forward diode bias v_D. Because this barrier is located in the oppositely doped material, the injected carriers at the barrier position are now minority carriers. As recombination takes hold, the minority carrier densities drop with depth to their equilibrium values for bulk minority carriers, a factor $\exp(-\varphi_B/V_{th})$ smaller than their bulk densities n_B, p_B as majority carriers before injection. At this point the quasi-Fermi levels rejoin the bulk Fermi level positions.

The reduced step in band edges also means that under forward bias the depletion region narrows as holes are pushed into it from the p-side and electrons from the n-side.

In the simple p–n diode the forward current increases exponentially with forward bias voltage due to the exponential increase in carrier densities, so there is always some current at even very small values of applied voltage. However, if one is interested in some particular current level, it will require a "knee" voltage before that current level is reached. For example, a very common choice in texts about circuits using silicon diodes is V_{Knee} = 0.7 V. Above the knee, the current continues to increase exponentially. Some special diodes, such as some varactors, are designed deliberately to maintain a low current level up to some knee voltage in the forward direction.

Reverse Bias

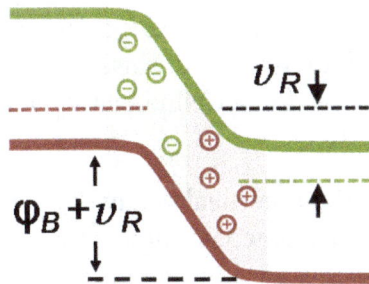

Band-bending for p–n diode in reverse bias

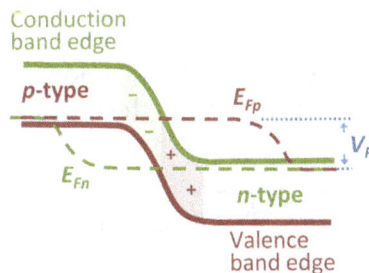

Quasi-Fermi levels in reverse-biased p–n diode.

In reverse bias the occupancy level for holes again tends to stay at the level of the bulk *p*-type semiconductor while the occupancy level for electrons follows that for the bulk *n*-type. In this case, the *p*-type bulk band edges are raised relative to the *n*-type bulk by the reverse bias v_R, so the two bulk occupancy levels are separated again by an energy determined by the applied voltage. As shown in the diagram, this behavior means the step in band edges is increased to $\varphi_B + v_R$, and the depletion region widens as holes are pulled away from it on the *p*-side and electrons on the *n*-side.

When the reverse bias is applied, the electric field in the depletion region is increased, pulling the electrons and holes further apart than in the zero bias case. Thus, any current that flows is due to the very weak process of carrier generation inside the depletion region due to *generation-recombination defects* in this region. That very small current is the source of the leakage current under reverse bias. In the photodiode, reverse current is introduced using creation of holes and electrons in the depletion region by incident light, thus converting a portion of the incident light into an electric current.

When the reverse bias becomes very large, reaching the breakdown voltage, the generation process in the depletion region accelerates leading to an *avalanche* condition which can cause runaway and destroy the diode.

Diode Law

The DC current-voltage behavior of the ideal *p*–*n* diode is governed by the Shockley diode equation:

$$i_D = I_R \left(e^{v_D / V_{th}} - 1 \right),$$

where v_D is the DC voltage across the diode and I_R is the *reverse saturation current*, the current that flows when the diode is reverse biased (that is, v_D is large and negative). The quantity V_{th} is the *thermal voltage* defined as $V_{th} = k_B T / q$. This is approximately equal to 25 mV at $T = 290$ kelvins.

This equation does not model the non-ideal behavior such as excess reverse leakage or breakdown phenomena. In many practical diodes this equation must be modified to read

$$i_D = I_R \left(e^{v_D / n V_{th}} - 1 \right),$$

where *n* is an *ideality factor* introduced to model a slower rate of increase than predicted by the ideal diode law. Using this equation, the diode *on*-resistance is

$$r_D = \frac{1}{di_D / dv_D} \approx \frac{n V_{th}}{i_D},$$

exhibiting a lower resistance the higher the current.

Capacitance

The depletion layer between the n- and p-sides of a $p-n$-diode serves as an insulating region that separates the two diode contacts. Thus, the diode in reverse bias exhibits a *depletion-layer capacitance*, sometimes more vaguely called a *junction capacitance*, analogous to a parallel plate capacitor with a dielectric spacer between the contacts. In reverse bias the width of the depletion layer is widened with increasing reverse bias v_R, and the capacitance is accordingly decreased. Thus, the junction serves as a voltage-controllable capacitor. In a simplified one-dimensional model, the junction capacitance is

$$C_J = \kappa\varepsilon_0 \frac{A}{w(v_R)},$$

with A the device area, κ the relative semiconductor dielectric permittivity, ε_o the electric constant, and w the depletion width (thickness of the region where mobile carrier density is negligible).

In forward bias, besides the above depletion-layer capacitance, minority carrier charge injection and diffusion occurs. A *diffusion capacitance* exists expressing the change in minority carrier charge that occurs with a change in forward bias. In terms of the stored minority carrier charge, the diode current i_D is

$$i_D = \frac{Q_D}{\tau_T},$$

where Q_D is the charge associated with diffusion of minority carriers, and τ_T is the *transit time*, the time taken for the minority charge to transit the injection region. Typical values for transit time are 0.1–100 ns. On this basis, the diffusion capacitance is calculated to be

$$C_D = \frac{dQ_D}{dv_D} = \tau_T \frac{di_D}{dv_D} = \frac{i_D \tau_T}{V_{th}}.$$

Generally speaking, for usual current levels in forward bias, this capacitance far exceeds the depletion-layer capacitance.

Transient Response

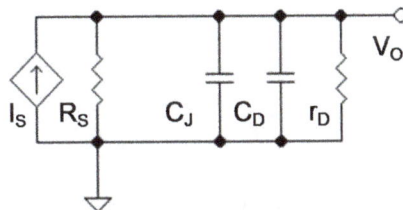

Small-signal circuit for $p-n$ diode driven by a current signal represented as a Norton source.

The diode is a highly non-linear device, but for small-signal variations its response can be analyzed using a *small-signal circuit* based upon the DC bias about which the signal is imagined to vary. The equivalent circuit is shown at the right for a diode driven by a Norton source. Using Kirchhoff's current law at the output node:

$$I_S = \left(j\omega(C_J + C_D) + \frac{1}{r_D} + \frac{1}{R_S} \right) V_O \,,$$

with C_D the diode diffusion capacitance, C_J the diode junction capacitance (the depletion layer capacitance) and r_D the diode resistance, all at the selected quiescent bias point or Q-point. The output voltage provided by this circuit is then:

$$\frac{V_O}{I_S} = \frac{(R_S \| r_D)}{1 + j\omega(C_D + C_J)(R_S \| r_D)} \,,$$

with $(R_S \| r_D)$ the parallel combination of R_S and r_D. This *transresistance amplifier* exhibits a *corner frequency*, denoted f_c:

$$f_C = \frac{1}{2\pi(C_D + C_J)(R_S \| r_D)} \,,$$

and for frequencies $f \gg f_c$ the gain rolls off with frequency as the capacitors short-circuit the resistor r_D. Assuming, as is the case when the diode is turned on, that $C_D \gg C_J$ and $R_S \gg r_D$, the expressions found above for the diode resistance and capacitance provide:

$$f_C = \frac{1}{2\pi n\tau_T} \,,$$

which relates the corner frequency to the diode transit time τ_T.

For diodes operated in reverse bias, C_D is zero and the term *corner frequency* often is replaced by *cutoff frequency*. In any event, in reverse bias the diode resistance becomes quite large, although not infinite as the ideal diode law suggests, and the assumption that it is less than the Norton resistance of the driver may not be accurate. The junction capacitance is small and depends upon the reverse bias v_R. The cutoff frequency is then:

$$f_C = \frac{1}{2\pi C_J (R_S \| r_D)} \,,$$

and varies with reverse bias because the width $w(v_R)$ of the insulating region depleted of mobile carriers increases with increasing diode reverse bias, reducing the capacitance.

p–n junction

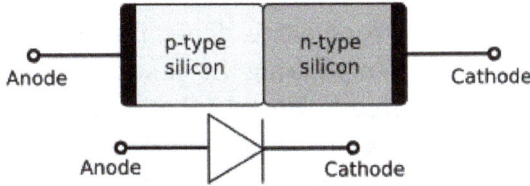

A p–n junction. The circuit symbol is shown: the triangle corresponds to the p side.

A p–n junction is a boundary or interface between two types of semiconductor material, p-type and n-type, inside a single crystal of semiconductor. The "p" (positive) side contains an excess of holes, while the "n" (negative) side contains an excess of electrons. The p-n junction is created by doping, for example by ion implantation, diffusion of dopants, or by epitaxy (growing a layer of crystal doped with one type of dopant on top of a layer of crystal doped with another type of dopant). If two separate pieces of material were used, this would introduce a grain boundary between the semiconductors that would severely inhibit its utility by scattering the electrons and holes.

p–n junctions are elementary "building blocks" of most semiconductor electronic devices such as diodes, transistors, solar cells, LEDs, and integrated circuits; they are the active sites where the electronic action of the device takes place. For example, a common type of transistor, the bipolar junction transistor, consists of two p–n junctions in series, in the form n–p–n or p–n–p.

The invention of the p–n junction is usually attributed to American physicist Russell Ohl of Bell Laboratories. However, Vadim Lashkaryov reported discovery of *p-n*-junctions in Cu_2O and silver sulphide photocells and selenium rectifiers in 1941.

A Schottky junction is a special case of a p–n junction, where metal serves the role of the p-type semiconductor.

Properties

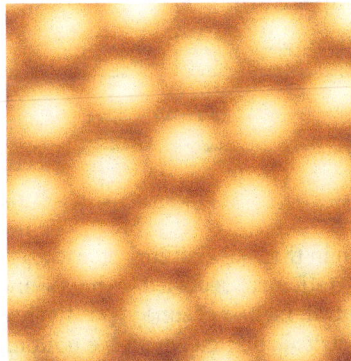

Image silicon atoms (Si) enlarged about 45,000,000x.

The p–n junction possesses some interesting properties that have useful applications in modern electronics. A p-doped semiconductor is relatively conductive. The same is true of an n-doped semiconductor, but the junction between them can become depleted of charge carriers, and hence non-conductive, depending on the relative voltages of the two semiconductor regions. By manipulating this non-conductive layer, p–n junctions are commonly used as diodes: circuit elements that allow a flow of electricity in one direction but not in the other (opposite) direction. *Bias* is the application of a voltage across a p–n junction; *forward bias* is in the direction of easy current flow, and *reverse bias* is in the direction of little or no current flow.

Equilibrium (Zero Bias)

In a p–n junction, without an external applied voltage, an equilibrium condition is reached in which a potential difference is formed across the junction. This potential difference is called built-in potential V_{bi}.

After joining p-type and n-type semiconductors, electrons from the n region near the p–n interface tend to diffuse into the p region leaving behind positively charged ions in the n region and being recombined with holes, forming negatively charged ions in the p region. Likewise, holes from the p-type region near the p–n interface begin to diffuse into the n-type region, leaving behind negatively charged ions in the p region and recombining with electrons, forming positive ions in the n region Template:Explain holes move?. The regions near the p–n interface lose their neutrality and most of their mobile carriers, forming the space charge region or depletion layer (see figure A).

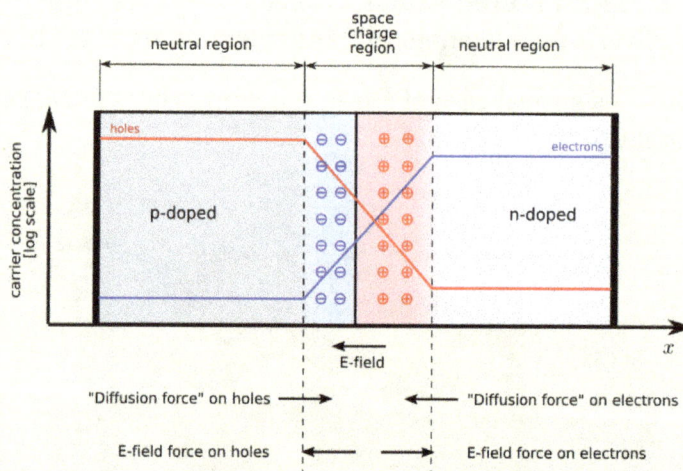

Figure A. A p–n junction in thermal equilibrium with zero-bias voltage applied. Electron and hole concentration are reported with blue and red lines, respectively. Gray regions are charge-neutral. Light-red zone is positively charged. Light-blue zone is negatively charged. The electric field is shown on the bottom, the electrostatic force on electrons and holes and the direction in which the diffusion tends to move electrons and holes. (The log concentration curves should actually be smoother with slope varying with field strength.)

The electric field created by the space charge region opposes the diffusion process for both electrons and holes. There are two concurrent phenomena: the diffusion process that tends to generate more space charge, and the electric field generated by the space charge that tends to counteract the diffusion. The carrier concentration profile at equilibrium is shown in figure A with blue and red lines. Also shown are the two counterbalancing phenomena that establish equilibrium.

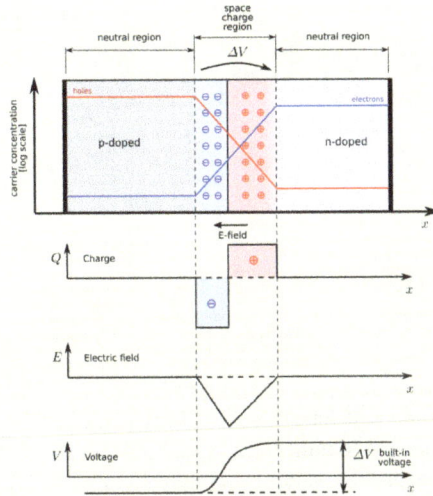

Figure B. A p–n junction in thermal equilibrium with zero-bias voltage applied. Under the junction, plots for the charge density, the electric field, and the voltage are reported. (The log concentration curves should actually be smoother, like the voltage.)

The space charge region is a zone with a net charge provided by the fixed ions (donors or acceptors) that have been left *uncovered* by majority carrier diffusion. When equilibrium is reached, the charge density is approximated by the displayed step function. In fact, since the y-axis of figure A is log-scale, the region is almost completely depleted of majority carriers (leaving a charge density equal to the net doping level), and the edge between the space charge region and the neutral region is quite sharp (see figure B, Q(x) graph). The space charge region has the same magnitude of charge on both sides of the p–n interfaces, thus it extends farther on the less doped side in this example (the n side in figures A and B).

Forward Bias

In forward bias, the p-type is connected with the positive terminal and the n-type is connected with the negative terminal.

With a battery connected this way, the holes in the p-type region and the electrons in the n-type region are pushed toward the junction and start to neutralize the depletion zone, reducing its width. The positive potential applied to the p-type material repels the holes, while the negative potential applied to the n-type material repels the electrons. The change in potential between the p side and the n side decreases or switches sign. With increasing forward-bias voltage, the depletion zone eventually becomes thin enough that the zone's electric field cannot counteract charge carrier motion across the p–n junction, which as a consequence reduces electrical resistance. The electrons that cross the p–n junction into the p-type material (or holes that cross into the n-type material) will diffuse into the nearby neutral region. The amount of minority diffusion in the near-neutral zones determines the amount of current that may flow through the diode.

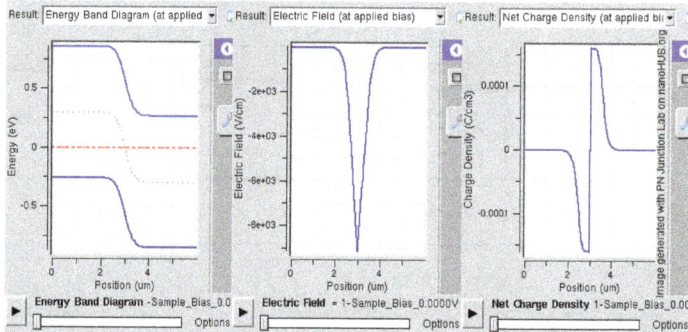

PN junction operation in forward-bias mode, showing reducing depletion width. The panels show energy band diagram, electric field, and net charge density. Both p and n junctions are doped at a 1e15/cm3 (0.00016C/cm³) doping level, leading to built-in potential of ~0.59 V. Reducing depletion width can be inferred from the shrinking charge profile, as fewer dopants are exposed with increasing forward bias. Observe the different quasi-fermi levels for conduction band and valence band in n and p regions (red curves)

Only majority carriers (electrons in n-type material or holes in p-type) can flow through a semiconductor for a macroscopic length. With this in mind, consider the flow of electrons across the junction. The forward bias causes a force on the electrons pushing them from the N side toward the P side. With forward bias, the depletion region is narrow enough that electrons can cross the junction and *inject* into the p-type material. However, they do not continue to flow through the p-type material indefinitely, because it is energetically favorable for them to recombine with holes. The average length an electron travels through the p-type material before recombining is called the *diffusion length*, and it is typically on the order of micrometers.

Although the electrons penetrate only a short distance into the p-type material, the electric current continues uninterrupted, because holes (the majority carriers) begin to flow in the opposite direction. The total current (the sum of the electron and hole currents) is constant in space, because any variation would cause charge buildup over time (this is Kirchhoff's current law). The flow of holes from the p-type region into the n-type region is exactly analogous to the flow of electrons from N to P (electrons and holes swap roles and the signs of all currents and voltages are reversed).

Therefore, the macroscopic picture of the current flow through the diode involves electrons flowing through the n-type region toward the junction, holes flowing through the p-type region in the opposite direction toward the junction, and the two species of carriers constantly recombining in the vicinity of the junction. The electrons and holes travel in opposite directions, but they also have opposite charges, so the overall current is in the same direction on both sides of the diode, as required.

The Shockley diode equation models the forward-bias operational characteristics of a p–n junction outside the avalanche (reverse-biased conducting) region.

Reverse Bias

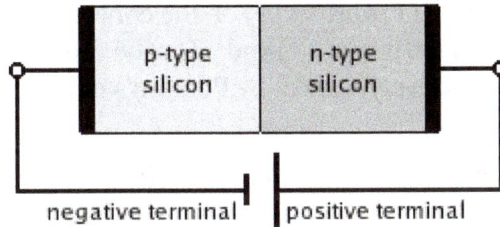

A silicon p–n junction in reverse bias.

Connecting the *p-type* region to the *negative* terminal of the battery and the *n-type* region to the *positive* terminal corresponds to reverse bias. If a diode is reverse-biased, the voltage at the cathode is comparatively higher than at the anode. Therefore, very little current will flow until the diode breaks down. The connections are illustrated in the adjacent diagram.

Because the p-type material is now connected to the negative terminal of the power supply, the 'holes' in the p-type material are pulled away from the junction, leaving behind charged ions and causing the width of the depletion region to increase. Likewise, because the n-type region is connected to the positive terminal, the electrons will also be pulled away from the junction, with similar effect. This increases the voltage barrier causing a high resistance to the flow of charge carriers, thus allowing minimal electric current to cross the p–n junction. The increase in resistance of the p–n junction results in the junction behaving as an insulator.

The strength of the depletion zone electric field increases as the reverse-bias voltage increases. Once the electric field intensity increases beyond a critical level, the p–n junction depletion zone breaks down and current begins to flow, usually by either the Zener or the avalanche breakdown processes. Both of these breakdown processes are non-destructive and are reversible, as long as the amount of current flowing does not reach levels that cause the semiconductor material to overheat and cause thermal damage.

This effect is used to advantage in Zener diode regulator circuits. Zener diodes have a low breakdown voltage. A standard value for breakdown voltage is for instance 5.6 V. This means that the voltage at the cathode cannot be more than about 5.6 V higher than the voltage at the anode (although there is a slight rise with current), because the diode will break down – and therefore conduct – if the voltage gets any higher. This in effect limits the voltage over the diode.

Another application of reverse biasing is Varicap diodes, where the width of the depletion zone (controlled with the reverse bias voltage) changes the capacitance of the diode.

Governing Equations

Size of Depletion Region

For a p–n junction, letting $C_A(x)$ and $C_D(x)$ be the concentrations of acceptor and donor atoms respectively, and letting $N_0(x)$ and $P_0(x)$ be the equilibrium concentrations of electrons and holes respectively, yields, by Poisson's equation:

$$-\frac{d^2V}{dx^2} = \frac{\rho}{\varepsilon} = \frac{q}{\varepsilon}\left[(N_0 - P_0) + (C_D - C_A)\right]$$

where V is the electric potential, ρ is the charge density, ε is permittivity and q is the magnitude of the electron charge. Letting d_p be the width of the depletion region within the p-side, and letting d_n be the width of the depletion region within the n-side, it must be that

$$d_p C_A = d_n C_D$$

because the total charge on either side of the depletion region must cancel out. Therefore, letting D and ΔV represent the entire depletion region and the potential difference across it,

$$\Delta V = \int_D \int \frac{q}{\varepsilon}\left[(N_0 - P_0) + (C_D - C_A)\right] dx dx$$

$$= \frac{C_A C_D}{C_A + C_D} \frac{q}{2\varepsilon}(d_p + d_n)^2$$

where $P_0 = N_0 = 0$, because we are in the depletion region. And thus, letting d be the total width of the depletion region, we get

$$d = \sqrt{\frac{2\varepsilon}{q} \frac{C_A + C_D}{C_A C_D} \Delta V}$$

ΔV can be written as $\Delta V_0 + \Delta V_{ext}$, where we have broken up the voltage difference into the equilibrium plus external components. The equilibrium potential results from diffusion forces, and thus we can calculate ΔV_0 by implementing the Einstein relation and assuming the semiconductor is nondegenerate (i.e. the product $P_0 N_0$ is independent of the Fermi energy)

$$\Delta V_0 = \frac{kT}{q} \ln\left(\frac{C_A C_D}{P_0 N_0}\right)$$

where T is the temperature of the semiconductor and k is Boltzmann constant.

Current Across Depletion Region

The *Shockley ideal diode equation* characterizes the current across a p–n junction as a function of external voltage and ambient conditions (temperature, choice of semiconductor, etc.). To see how it can be derived, we must examine the various reasons for current. The convention is that the forward (+) direction be pointed against the diode's built-in potential gradient at equilibrium.

- Forward Current ($\mathbf{J}_F$)

    o Diffusion Current: current due to local imbalances in carrier concentration n, via the equation $\mathbf{J}_D \propto -q\nabla n$

- Reverse Current ($\mathbf{J}_R$)

    o Field Current

    o Generation Current

The forward-bias and the reverse-bias properties of the p–n junction imply that it can be used as a diode. A p–n junction diode allows electric charges to flow in one direction, but not in the opposite direction; negative charges (electrons) can easily flow through the junction from n to p but not from p to n, and the reverse is true for holes. When the p–n junction is forward-biased, electric charge flows freely due to reduced resistance of the p–n junction. When the p–n junction is reverse-biased, however, the junction barrier (and therefore resistance) becomes greater and charge flow is minimal.

Non-rectifying Junctions

In the above diagrams, contact between the metal wires and the semiconductor material also creates metal–semiconductor junctions called Schottky diodes. In a simplified ideal situation a semiconductor diode would never function, since it would be composed of several diodes connected back-to-front in series. But, in practice, surface impurities within the part of the semiconductor that touches the metal terminals will greatly reduce the width of those depletion layers to such an extent that the metal-semiconductor junctions do not act as diodes. These *non-rectifying junctions* behave as ohmic contacts regardless of applied voltage polarity.

p-n Junction Diode

Diode

A pure silicon crystal or germanium crystal is known as an intrinsic semiconductor. There are not enough free electrons and holes in an intrinsic semi-conductor to produce a usable current. The electrical action of these can be modified by doping means adding impurity atoms to a crystal to increase either the number of free holes or no of free electrons.

When a crystal has been doped, it is called a extrinsic semi-conductor. They are of two types

- n-type semiconductor having free electrons as majority carriers

- p-type semiconductor having free holes as majority carriers

By themselves, these doped materials are of little use. However, if a junction is made by joining p-type semiconductor to n-type semiconductor a useful device is produced known as diode. It will allow current to flow through it only in one direction. The unidirectional properties of a diode allow current flow when forward biased and disallow current flow when reversed biased. This is called rectification process and therefore it is also called rectifier.

How is it possible that by properly joining two semiconductors each of which, by itself, will freely conduct the current in any direct refuses to allow conduction in one direction.

Consider first the condition of p-type and n-type germanium just prior to joining in the figure below. The majority and minority carriers are in constant motion.

The minority carriers are thermally produced and they exist only for short time after which they recombine and neutralize each other. In the mean time, other minority carriers have been produced and this process goes on and on.

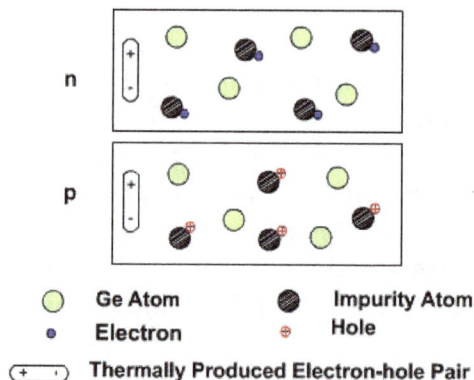

The number of these electron hole pair that exist at any one time depends upon the temperature. The number of majority carriers is however, fixed depending on the number of impurity atoms available. While the electrons and holes are in motion but the atoms are fixed in place and do not move.

As soon as, the junction is formed, the following processes are initiated.

- Holes from the p-side diffuse into n-side where they recombine with free electrons.

- Free electrons from n-side diffuse into p-side where they recombine with free holes.

- The diffusion of electrons and holes is due to the fact that large no of electrons are concentrated in one area and large no of holes are concentrated in another area.

- When these electrons and holes begin to diffuse across the junction then they collide each other and negative charge in the electrons cancels the positive charge of the hole and both will lose their charges.

- The diffusion of holes and electrons is an electric current referred to as a recombination current. The recombination process decay exponentially with both time and distance from the junction. Thus most of the recombination occurs just after the junction is made and very near to junction.

- A measure of the rate of recombination is the lifetime defined as the time required for the density of carriers to decrease to 37% to the original concentration

The impurity atoms are fixed in their individual places. The atoms itself is a part of the crystal and so cannot move. When the electrons and hole meet, their individual charge is cancelled and this leaves the originating impurity atoms with a net charge, the atom that produced the electron now lack an electronic and so becomes charged positively, whereas the atoms that produced the hole now lacks a positive charge and becomes negative.

The electrically charged atoms are called ions since they are no longer neutral. These ions produce an electric field. After several collisions occur, the electric field is great enough to repel rest of the majority carriers away of the junction. For example, an electron trying to diffuse from n to p side is repelled by the negative charge of the p-side. Thus diffusion process does not continue indefinitely but continues as long as the field is developed.

This region is produced immediately surrounding the junction that has no majority carriers. The majority carriers have been repelled away from the junction and junction is depleted from carriers. The junction is known as the barrier region or depletion region. The electric field represents a potential difference across the junction also called *space charge potential or barrier potential* . This potential is 0.7v for Si at 25° celcious and 0.3v for Ge.

The physical width of the depletion region depends on the doping level. If very heavy doping is used, the depletion region is physically thin because diffusion charge need not travel far across the junction before recombination takes place (short life time). If doping is light, then depletion is more wide (long life time).

The symbol of diode is shown in the figure. The terminal connected to p-layer is called anode (A) and the terminal connected to n-layer is called cathode (K)

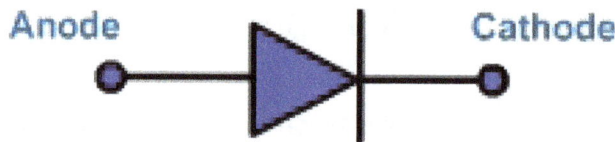

Anode   Cathode

Reverse Bias

If positive terminal of dc source is connected to cathode and negative terminal is connected to anode, the diode is called reverse biased as shown in the figure below.

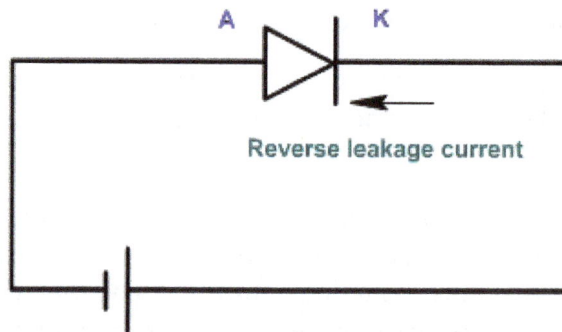

A   K

Reverse leakage current

When the diode is reverse biased then the depletion region width increases, majority carriers move away from the junction and there is no flow of current due to majority carriers but there are thermally produced electron hole pair also. If these electrons and holes are generated in the vicinity of junction then there is a flow of current. The negative voltage applied to the diode will tend to attract the holes thus generated and repel the electrons. At the same time, the positive voltage will attract the electrons towards the battery and repel the holes. This will cause current to flow in the circuit. This current is usually very small (interms of micro amp to nano amp). Since this current is due to minority carriers and these number of minority carriers are fixed at a given temperature therefore, the current is almost constant known as reverse saturation current I_{co}.

In actual diode, the current is not almost constant but increases slightly with voltage. This is due to surface leakage current. The surface of diode follows ohmic law (V=IR). The resistance under reverse bias condition is very high 100k to mega ohms. When the reverse voltage is increased, then at certain voltage, then breakdown to diode takes place and it conducts heavily. This is due to avalanche or zener breakdown. The characteristic of the diode is shown in the following figure.

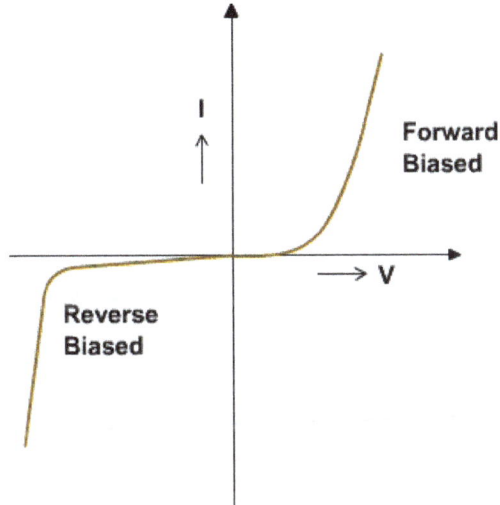

Forward Bias

When the diode is forward bias, then majority carriers are pushed towards junction, when they collide and recombination takes place. Number of majority carriers are fixed in semiconductor. Therefore as each electron is eliminated at the junction, a new electron must be introduced, this comes from battery. At the same time, one hole must be created in p-layer. This is formed by extracting one electron from p-layer. Therefore, there is a flow of carriers and thus flow of current.

Diode Operating Point

Example - 1

When a silicon diode is conducting at a temperature of 25°C, a 0.7 V drop exists across its terminals. What is the voltage, V_{ON}, across the diode at 100°C?

Solution

The temperature relationship is described by

$$V_{ON}(T_{New}) - V_{ON}(T_{room}) = K_T (T_{New} - T_{room})$$

or, $$V_{ON}(T_{New}) = V_{ON}(T_{room}) + K_T (T_{new} - T_{room})$$

Given $V_{ON}(T_{room}) = 0,7$ V, $T_{room} = 25°$ C, $T_{New} = 100°$ C

Therefore, $V_{ON}(T_{New}) = 0.7 + (-2 \times 10^{-3})(100-75) = 0.55$ V

Example - 2

Find the output current for the circuit shown in fig. (a).

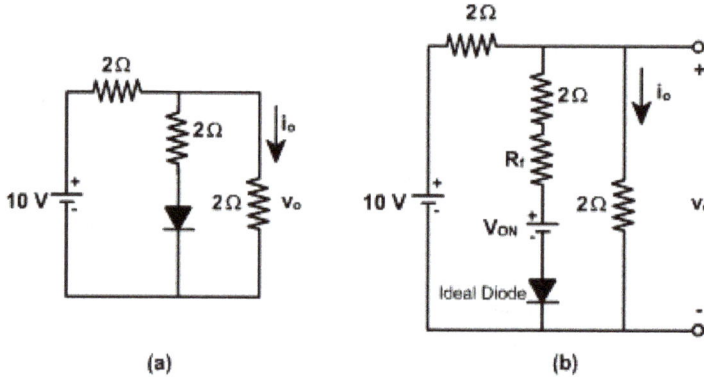

(a)

(b)

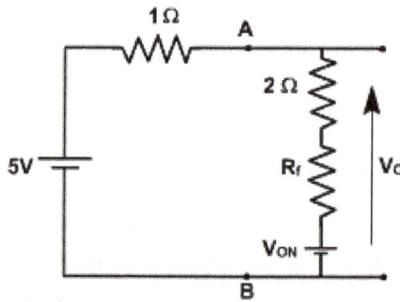

(C)

Circuit for Example 2

Solutions

Since the problem contains only a dc source, we use the diode equivalent circuit, as shown in fig. (b). Once we determine the state of the ideal diode in this model (i.e., either open circuit or short circuit), the problem becomes one of simple dc circuit analysis.

It is reasonable to assume that the diode is forward biased. This is true since the only external source is 10 V, which clearly exceeds the turn-on voltage of the diode, even taking the voltage division into account. The equivalent circuit then becomes that of fig. (b). with the diode replaced by a short circuit.

The Thevenin's equivalent of the circuit between A and B is given by fig. (c).

The output voltage is given by

$$v_o = \left(\frac{5-V_{on}}{3+R_f}\right)(2+R_f)+V_{on}$$

or, $v_o = \dfrac{10+V_{ON}+5R_f}{3+R_f}$

If $V_{ON} = 0.7V$, and $R_f = 0.2\ W$, then

$V_o = 3.66V$

Example - 3

The circuit in the figure below, has a source voltage of $V_s = 1.1 + 0.1 \sin 1000t$. Find the current, i_D. Assume that

$nV_T = 40\ mV$

$V_{ON} = 0.7\ V$

Solution

We use KVL for dc equation to yield

$V_s = V_{ON} + l_D\ R_L$

$$l_D = \frac{V_s - V_{ON}}{R_L} = 4\ mA$$

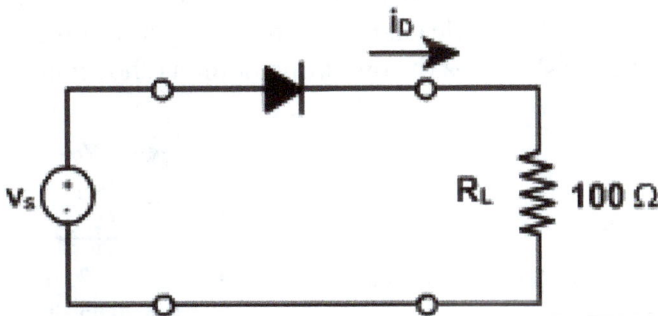

This sets the dc operating point of the diode. We need to determine the dynamic resistance so we can establish the resistance of the forward-biased junction for the ac signal.

$$r_D = \frac{nV_T}{l_D} = 10\ \Omega$$

Assuming that the contact resistance is negligible $R_f = r_D$ Now we can replace the forward-biased diode with a 10 W resistor. Again using KVL, we have,

$$v_s = R_f i_d + R_L i_d$$

$$i_d = \frac{v_S}{R_f + R_L} = 0.91 \sin 1000\, t \; mA$$

The diode current is given by

I = 4 + 0.91 sin 1000 t mA

Since i_D is always positive, the diode is always forward-biased, and the solution is complete.

Small Signal Operation of Real Diode

Consider the diode circuit

$$V = V_D + I_d R_L$$

$$V_D = V - I_d R_L$$

This equation involves two unknowns and cannot be solved. The straight line represented by the above equation is known as the load line. The load line passes through two points,

$$I = 0, V_D = V$$

and $\qquad V_D = 0, I = V / R_L.$

The slope of this line is equal to $1/R_L$. The other equation in terms of these two variables V_D & I_d, is given by the static characteristic. The point of intersection of straight line and diode characteristic gives the operating point as shown in the left figure.

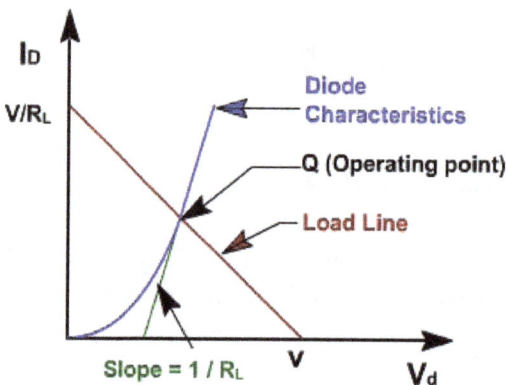

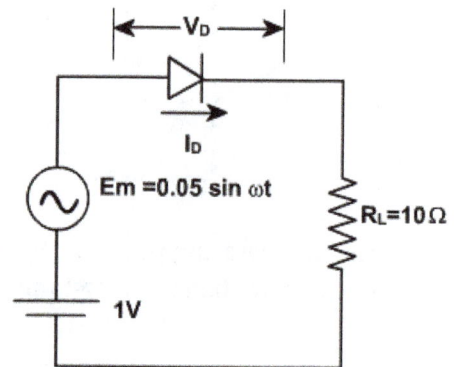

Let us consider a circuit shown in the right figure having dc voltage and sinusoidal ac voltage. Say V = 1V, R_L =10 ohm.

The resulting input voltage is the sum of dc voltage and sinusoidal ac voltage. There-fore, as the diode voltage varies, diode current also varies, sinusoidally. The intersec-tion of load line and diode characteristic for different input voltages gives the output voltage as shown in the figure below.

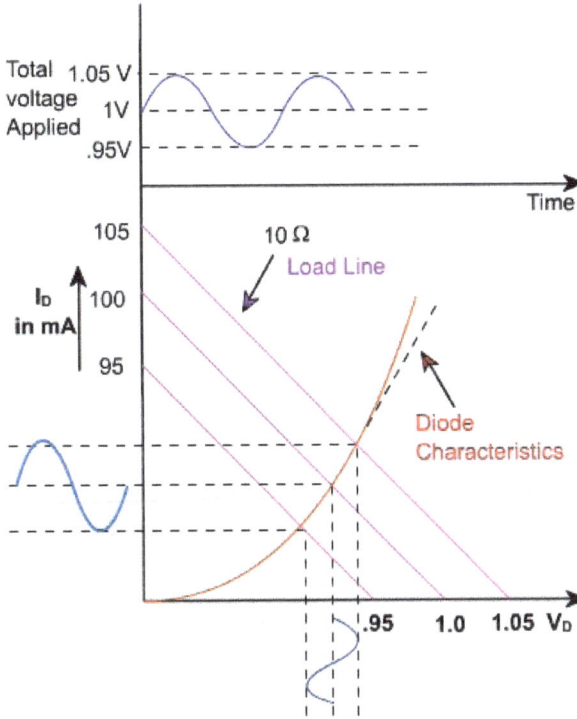

In certain applications only ac equivalent circuit is required. Since only ac response of the circuit is considered DC Source is not shown in the equivalent circuit of figure be-low. The resistance r_f represents the dynamic resistance or ac resistance of the diode. It is obtained by taking the ratio of $\Delta V_D / \Delta I_D$ at operating point.

Dynamic Resistance $\Delta r_D = \Delta V_D / \Delta I_D$

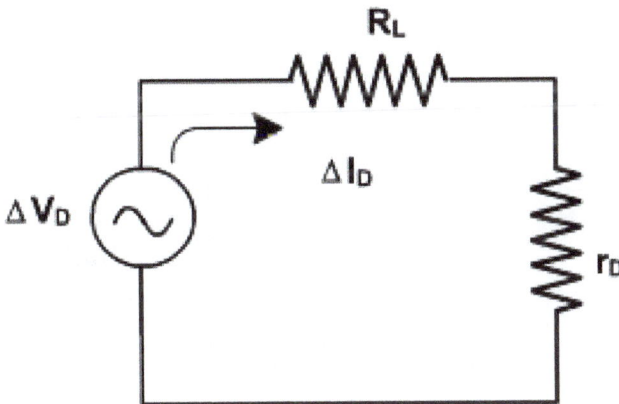

Applications of Diode

Diode Approximation: (Large Signal Operations)

Ideal Diode

- When diode is forward biased, resistance offered is zero,
- When it is reverse biased resistance offered is infinity. It acts as a perfect switch.

The characteristic and the equivalent circuit of the diode is shown in the figure.

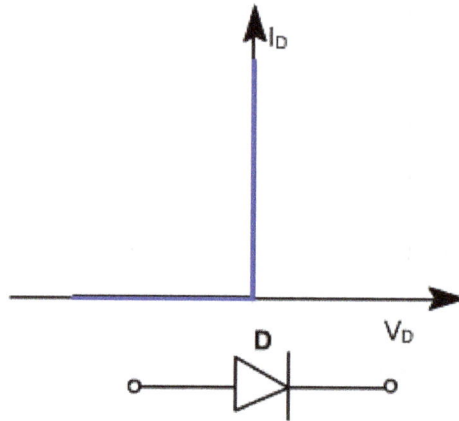

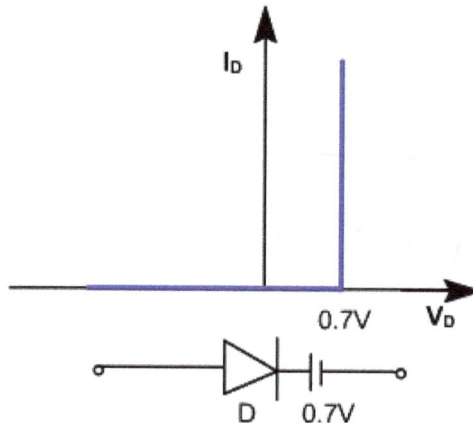

Second Approximation

- When forward voltage is more than 0.7 V, for Si diode then it conducts and offers zero resistance. The drop across the diode is 0.7V.
- When reverse biased it offers infinite resistance.

The characteristic and the equivalent circuit is shown in the figure.

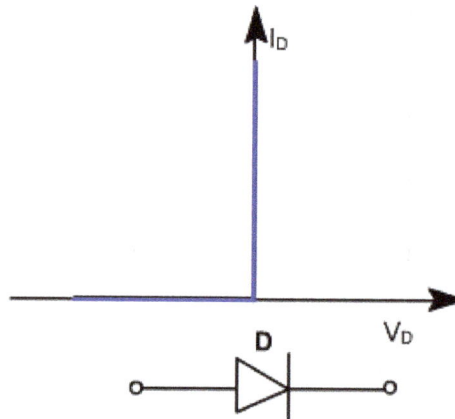

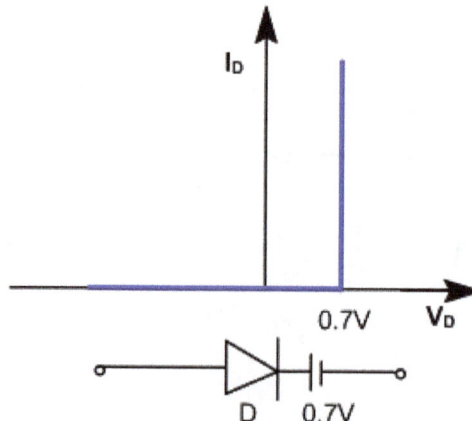

3rd Approximation

- When forward voltage is more than 0.7 V, then the diode conducts and the voltage drop across the diode becomes 0.7 V and it offers resistance R_f (slope of the current)

$V_D = 0.7 + I_D R_f$

The output characteristic and the equivalent circuit is shown in the figure.

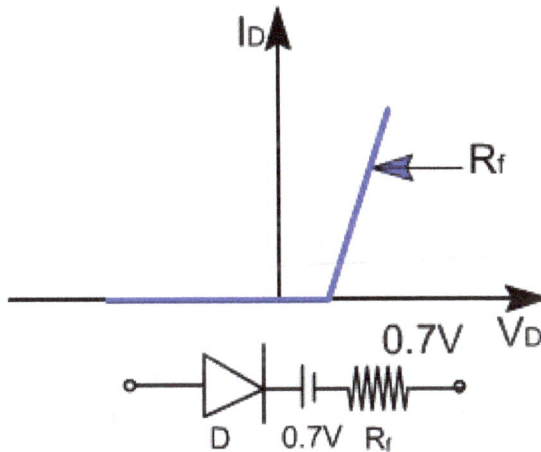

- When reverse biased resistance offered is very high & not infinity, then the diode equivalent circuit is as shown in the figure.

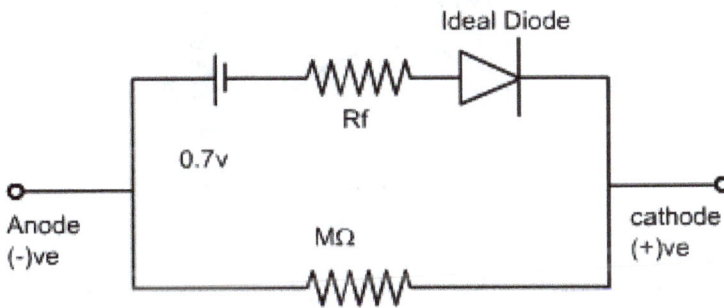

Example - 1

Calculate the voltage output of the circuit shown in the figure for following inputs

(a) $V_1 = V_2 = 0$.

(b) $V_1 = V, V_2 = 0$.

(c) $V_1 = V_2 = V$ knew voltage $= V_r$

Forward resistance of each diode is R_f.

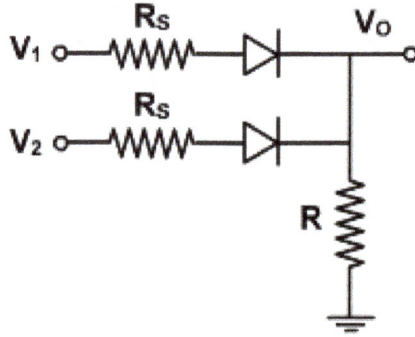

Solution

(a). When both V_1 and V_2 are zero , then the diodes are unbiased. Therefore,

$V_o = 0$ V

(b). When $V_1 = V$ and $V_2 = 0$, then one upper diode is forward biased and lower diode is unbiased. The resultant circuit using third approximation of diode will be as shown in the figure above.

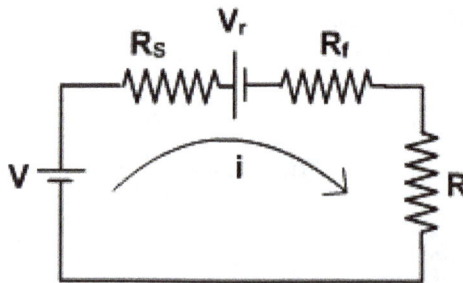

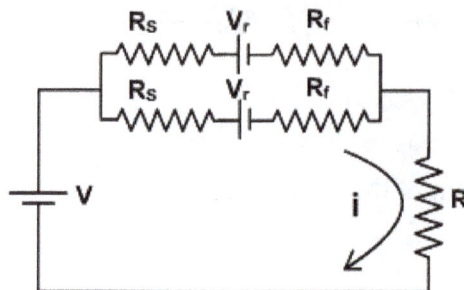

Applying KVL, we get

$$V = I(R_f + R_s + R) + V_r$$

$$\therefore I = \frac{V - Vr}{R_s - R_f + R}$$

(c) When both V_1 and V_2 are same as V, then both the diodes are forward biased and conduct. The resultant circuit using third approximation of diode will be as shown in the above figure.

$$V = \frac{1}{2}(R_f + R_s) + V_r + 1$$

$$I = \frac{V + V_r}{\left(\dfrac{R_s + R_f}{2} + R\right)}$$

Half Wave Rectifier

The single – phase half wave rectifier is shown in the figure.

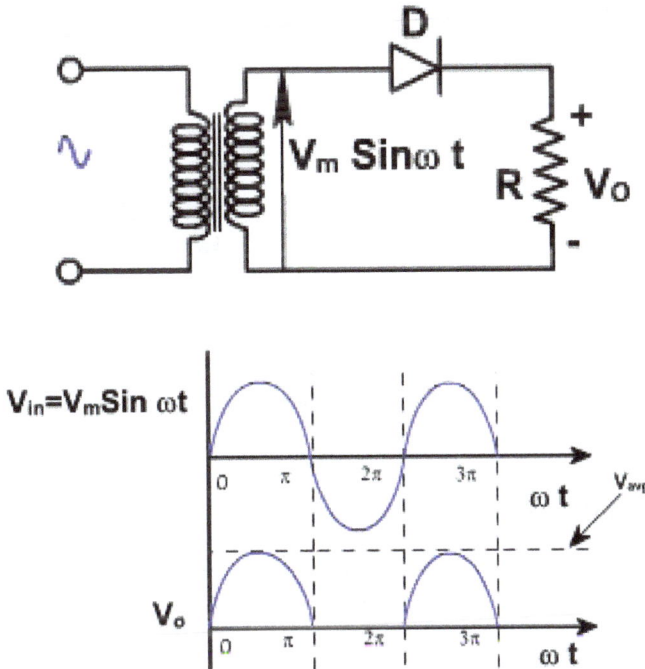

In positive half cycle, D is forward biased and conducts. Thus the output voltage is same as the input voltage. In the negative half cycle, D is reverse biased, and therefore output voltage is zero. The output voltage waveform is shown in the figure.

The average output voltage of the rectifier is given by

$$V_{avg} = \frac{1}{2} \int_0^z V_m \sin \omega t \, d(\omega t)$$

$$= \frac{V_m}{\pi} = 0.318 V_m$$

The average output current is given by

$$I_{avg} = \frac{V_m}{\pi R}$$

When the diode is reverse biased, entire transformer voltage appears across the diode. The maximum voltage across the diode is V_m. The diode must be capable to withstand this voltage. Therefore PIV half wave rating of diode should be equal to V_m in case of single-phase rectifiers. The average current rating must be greater than I_{avg}.

Full Wave Rectifier

A single – phase full wave rectifier using center tap transformer is shown in the figure. It supplies current in both half cycles of the input voltage.

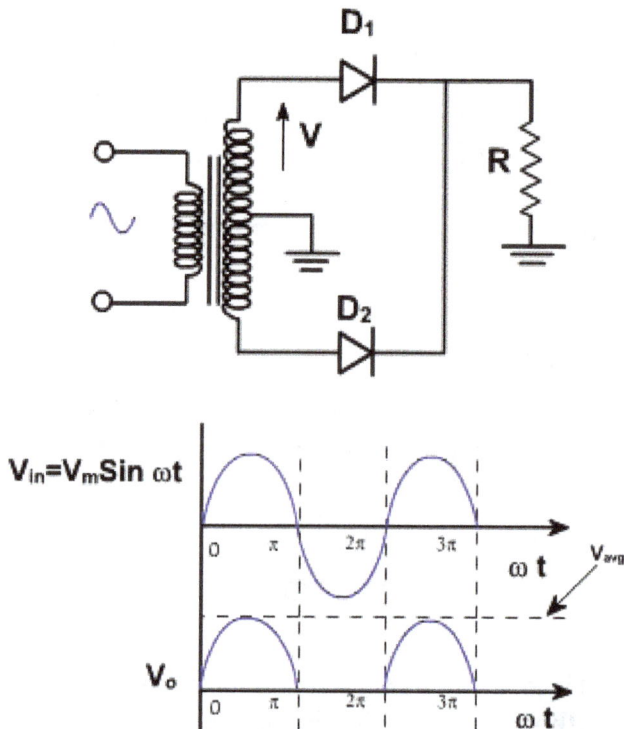

In the first half cycle D_1 is forward biased and conducts. But D_2 is reverse biased and does not conduct. In the second half cycle D_2 is forward biased, and conducts and D_1 is reverse biased. It is also called 2 – pulse midpoint converter because it supplies current in both the half cycles. The output voltage waveform is shown in the above figure.

The average output voltage is given by

$$V_{avg} = \frac{1}{2} \int_{0}^{z} V_m \sin \omega t \, d(\omega t)$$

$$= \frac{2V_m}{\pi}$$

and the average load current is given by

$$I_{avg} = \frac{2V_m}{\pi R}$$

When D_1 conducts, then full secondary voltage appears across D_2, therefore PIV rating of the diode should be 2 V_m.

Zener Diode

A Zener diode is a particular type of diode that, unlike a normal one, allows current to flow not only from its anode to its cathode, but also in the reverse direction, when the so-called "Zener voltage" is reached. Zener diodes have a highly doped p-n junction. Normal diodes will also break down with a reverse voltage but the voltage and sharpness of the knee are not as well defined as for a Zener diode. Also normal diodes are not designed to operate in the breakdown region, but Zener diodes can reliably operate in this region.

The device was named after Clarence Melvin Zener, who discovered the Zener effect. Zener reverse breakdown is due to electron quantum tunnelling caused by a high strength electric field. However, many diodes described as "Zener" diodes rely instead on avalanche breakdown. Both breakdown types are used in Zener diodes with the Zener effect predominating under 5.6 V and avalanche breakdown above.

Zener diodes are widely used in electronic equipment of all kinds and are one of the basic building blocks of electronic circuits. They are used to generate low power stabilized supply rails from a higher voltage and to provide reference voltages for circuits, especially stabilized power supplies. They are also used to protect circuits from over-voltage, especially electrostatic discharge (ESD).

Operation

A conventional solid-state diode allows significant current if it is reverse-biased above its reverse breakdown voltage. When the reverse bias breakdown voltage is exceeded, a conventional diode is subject to high current due to avalanche breakdown. Unless this current is limited by circuitry, the diode may be permanently damaged due to overheating. A Zener diode exhibits almost the same properties, except the device is specially designed so as to have a reduced breakdown voltage, the so-called Zener voltage. By contrast with the conventional device, a reverse-biased Zener diode exhibits a controlled breakdown and allows the current to keep the voltage across the Zener diode close to the Zener breakdown voltage. For example, a diode with a Zener breakdown voltage of 3.2 V exhibits a voltage drop of very nearly 3.2 V across a wide range of reverse currents. The Zener diode is therefore ideal for applications such as the generation of a reference voltage (e.g. for an amplifier stage), or as a voltage stabilizer for low-current applications.

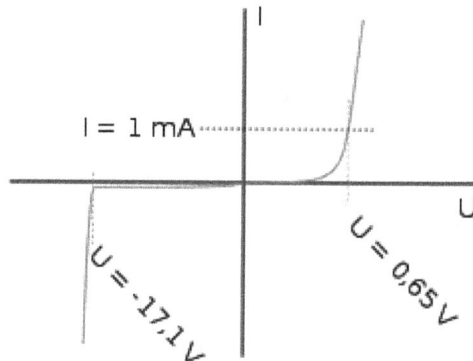

Current-voltage characteristic of a Zener diode with a breakdown voltage of 17 volts. Notice the change of voltage scale between the forward biased (positive) direction and the reverse biased (negative) direction.

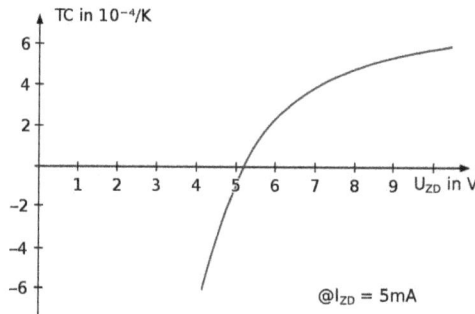

Temperature coefficient of Zener voltage against nominal Zener voltage.

Another mechanism that produces a similar effect is the avalanche effect as in the avalanche diode. The two types of diode are in fact constructed the same way and both effects are present in diodes of this type. In silicon diodes up to about 5.6 volts, the Zener effect is the predominant effect and shows a marked negative temperature coefficient. Above 5.6 volts, the avalanche effect becomes predominant and exhibits a positive temperature coefficient.

In a 5.6 V diode, the two effects occur together, and their temperature coefficients nearly cancel each other out, thus the 5.6 V diode is useful in temperature-critical applications. An alternative, which is used for voltage references that need to be highly stable over long periods of time, is to use a Zener diode with a temperature coefficient (TC) of +2 mV/°C (breakdown voltage 6.2–6.3 V) connected in series with a forward-biased silicon diode (or a transistor B-E junction) manufactured on the same chip. The forward-biased diode has a temperature coefficient of −2 mV/°C, causing the TCs to cancel out.

Modern manufacturing techniques have produced devices with voltages lower than 5.6 V with negligible temperature coefficients, but as higher-voltage devices are encountered, the temperature coefficient rises dramatically. A 75 V diode has 10 times the coefficient of a 12 V diode.

Zener and avalanche diodes, regardless of breakdown voltage, are usually marketed under the umbrella term of "Zener diode".

Under 5.6 V, where the Zener effect dominates, the IV curve near breakdown is much more rounded, which calls for more care in targeting its biasing conditions. The IV curve for Zeners above 5.6 V (being dominated by Avalanche), is much sharper at breakdown.

Waveform Clipper

Two Zener diodes facing each other in series will act to clip both halves of an input signal. Waveform clippers can be used to not only reshape a signal, but also to prevent voltage spikes from affecting circuits that are connected to the power supply.

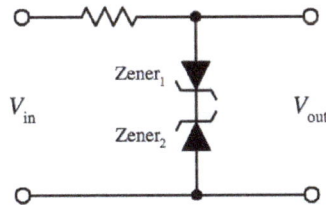

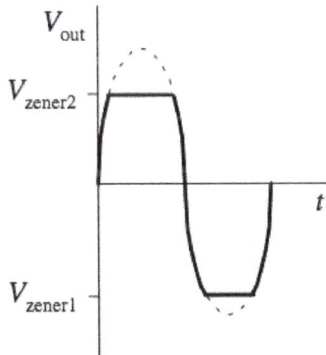

Examples of a Waveform Clipper

Voltage Shifter

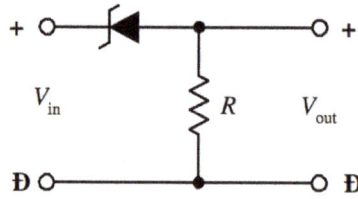

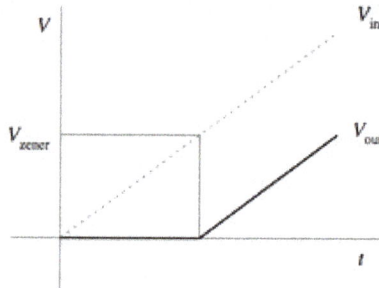

Examples of a Voltage Shifter

A Zener diode can be applied to a circuit with a resistor to act as a voltage shifter. This circuit lowers the output voltage by a quantity that is equal to the Zener diode's breakdown voltage.

Voltage Regulator

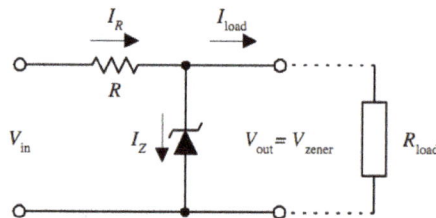

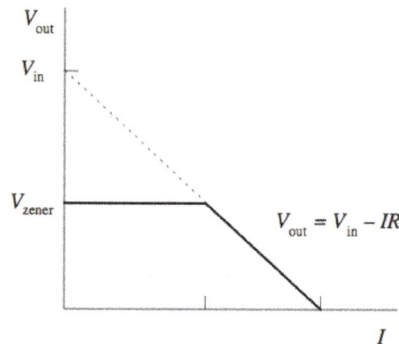

Examples of a Voltage Regulator

A Zener diode can be applied in a voltage regulator circuit to regulate the voltage applied to a load, such as in a linear regulator.

Construction

The Zener diode's operation depends on the heavy doping of its p-n junction. The depletion region formed in the diode is very thin (<1 μm) and the electric field is consequently very high (about 500 kV/m) even for a small reverse bias voltage of about 5 V, allowing electrons to tunnel from the valence band of the p-type material to the conduction band of the n-type material.

At the atomic scale, this tunnelling corresponds to the transport of valence band electrons into the empty conduction band states; as a result of the reduced barrier between these bands and high electric fields that are induced due to the relatively high levels of doping on both sides. The breakdown voltage can be controlled quite accurately in the doping process. While tolerances within 0.07% are available, the most widely used tolerances are 5% and 10%. Breakdown voltage for commonly available Zener diodes can vary widely from 1.2 volts to 200 volts.

For diodes that are lightly doped the breakdown is dominated by the avalanche effect rather than the Zener effect. Consequently, the breakdown voltage is higher (over 5.6 V) for these devices.

Surface Zeners

The emitter-base junction of a bipolar NPN transistor behaves as a Zener diode, with breakdown voltage at about 6.8 V for common bipolar processes and about 10 V for lightly doped base regions in BiCMOS processes. Older processes with poor control of doping characteristics had the variation of Zener voltage up to ±1 V, newer processes using ion implantation can achieve no more than ±0.25 V. The NPN transistor structure can be employed as a *surface Zener diode*, with collector and emitter connected together as its cathode and base region as anode. In this approach the base doping profile usually narrows towards the surface, creating a region with intensified electric field where the avalanche breakdown occurs. The hot carriers produced by acceleration in the intense field sometime shoot into the oxide layer above the junction and become trapped there. The accumulation of trapped charges can then cause 'Zener walkout', a corresponding change of the Zener voltage of the junction. The same effect can be achieved by radiation damage.

The emitter-base Zener diodes can handle only smaller currents as the energy is dissipated in the base depletion region which is very small. Higher amount of dissipated energy (higher current for longer time, or a short very high current spike) causes thermal damage to the junction and/or its contacts. Partial damage of the junction can shift its Zener voltage. Total destruction of the Zener junction by overheating it and causing migration of metallization across the junction ("spiking") can be used intentionally as a 'Zener zap' antifuse.

Subsurface Zeners

A subsurface Zener diode, also called 'buried Zener', is a device similar to the Surface Zener, but with the avalanche region located deeper in the structure, typically several micrometers below the oxide. The hot carriers then lose energy by collisions with the semiconductor lattice before reaching the oxide layer and cannot be trapped there. The Zener walkout phenomenon therefore does not occur here, and the buried Zeners have voltage constant over their entire lifetime. Most buried Zeners have breakdown voltage of 5–7 volts. Several different junction structures are used.

Uses

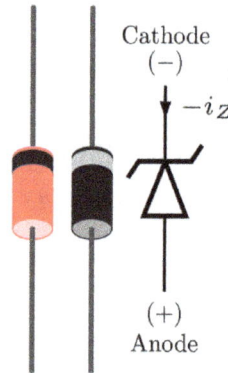

Zener diode shown with typical packages. *Reverse* current $-i_Z$ is shown.

Zener diodes are widely used as voltage references and as shunt regulators to regulate the voltage across small circuits. When connected in parallel with a variable voltage source so that it is reverse biased, a Zener diode conducts when the voltage reaches the diode's reverse breakdown voltage. From that point on, the relatively low impedance of the diode keeps the voltage across the diode at that value.

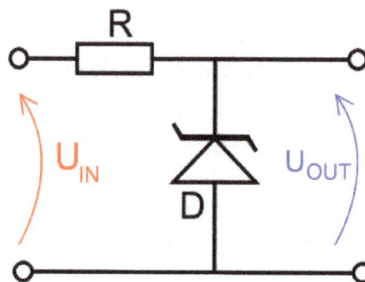

In this circuit, a typical voltage reference or regulator, an input voltage, U_{IN}, is regulated down to a stable output voltage U_{OUT}. The breakdown voltage of diode D is stable over a wide current range and holds U_{OUT} relatively constant even though the input voltage may fluctuate over a fairly wide range. Because of the low impedance of the diode when operated like this, resistor R is used to limit current through the circuit.

In the case of this simple reference, the current flowing in the diode is determined using Ohm's law and the known voltage drop across the resistor R;

$$I_{diode} = \frac{U_{IN} - U_{OUT}}{R_{\Omega}}$$

The value of R must satisfy two conditions :

1. R must be small enough that the current through D keeps D in reverse break-down. For example, the common BZX79C5V6 device, a 5.6 V 0.5 W Zener diode, has a recommended reverse current of 5 mA. If insufficient current exists through D, then U_{OUT} is unregulated and less than the nominal breakdown voltage (this differs to voltage-regulator tubes where the output voltage will be higher than nominal and could rise as high as U_{IN}). When calculating R, allowance must be made for any current through the external load, not shown in this diagram, connected across U_{OUT}.

2. R must be large enough that the current through D does not destroy the device. If the current through D is I_D, its breakdown voltage V_B and its maximum power dissipation P_{MAX} correlate as such: $I_D V_B < P_{MAX}$.

A load may be placed across the diode in this reference circuit, and as long as the Zener stays in reverse breakdown, the diode provides a stable voltage source to the load. Zener diodes in this configuration are often used as stable references for more advanced voltage regulator circuits.

Shunt regulators are simple, but the requirements that the ballast resistor be small enough to avoid excessive voltage drop during worst-case operation (low input voltage concurrent with high load current) tends to leave a lot of current flowing in the diode much of the time, making for a fairly wasteful regulator with high quiescent power dissipation, only suitable for smaller loads.

These devices are also encountered, typically in series with a base-emitter junction, in transistor stages where selective choice of a device centered around the avalanche or Zener point can be used to introduce compensating temperature co-efficient balancing of the transistor p–n junction. An example of this kind of use would be a DC error amplifier used in a regulated power supply circuit feedback loop system.

Zener diodes are also used in surge protectors to limit transient voltage spikes.

Another application of the Zener diode is the use of noise caused by its avalanche breakdown in a random number generator.

The diodes designed to work in breakdown region are called zener diode. If the reverse voltage exceeds the breakdown voltage, the zener diode will normally not be destroyed as long as the current does not exceed maximum value and the device closes not over load.

When a thermally generated carrier (part of the reverse saturation current) falls down the junction and acquires energy of the applied potential, the carrier collides with crystal ions and imparts sufficient energy to disrupt a covalent bond. In addition to the original carrier, a new electron-hole pair is generated. This pair may pick up sufficient energy from the applied field to collide with another crystal ion and create still another electron-hole pair. This action continues and thereby disrupts the covalent bonds. The process is referred to as impact ionization, avalanche multiplication or avalanche breakdown.

There is a second mechanism that disrupts the covalent bonds. The use of a sufficiently strong electric field at the junction can cause a direct rupture of the bond. If the electric field exerts a strong force on a bound electron, the electron can be torn from the covalent bond thus causing the number of electron-hole pair combinations to multiply. This mechanism is called high field emission or Zener breakdown. The value of reverse voltage at which this occurs is controlled by the amount ot doping of the diode. A heavily doped diode has a low Zener breakdown voltage, while a lightly doped diode has a high Zener breakdown voltage.

At voltages above approximately 8V, the predominant mechanism is the avalanche breakdown. Since the Zener effect (avalanche) occurs at a predictable point, the diode can be used as a voltage reference. The reverse voltage at which the avalanche occurs is called the breakdown or Zener voltage.

A typical Zener diode characteristic is shown in figure. The circuit symbol for the Zener diode is different from that of a regular diode, and is illustrated in the figure. The maximum reverse current, $I_{Z(max)}$, which the Zener diode can withstand is dependent on the design and construction of the diode. A design guideline that the minimum Zener current, where the characteristic curve remains at V_Z (near the knee of the curve), is $0.1/ I_{Z(max)}$.

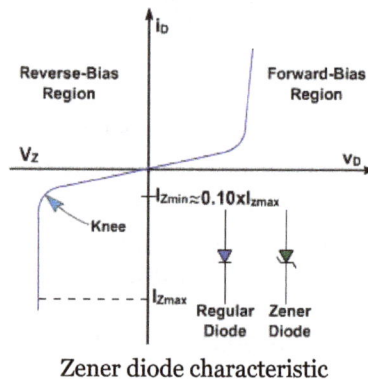

Zener diode characteristic

The power handling capacity of these diodes is better. The power dissipation of a zener diode equals the product of its voltage and current.

$$P_z = V_z I_z$$

The amount of power which the zener diode can withstand ($V_z.I_{Z(max)}$) is a limiting factor in power supply design.

Zener Regulator

When zener diode is forward biased it works as a diode and drop across it is 0.7 V. When it works in breakdown region the voltage across it is constant (V_z) and the current through diode is decided by the external resistance. Thus, zener diode can be used as a voltage regulator in the configuration shown in the left figure for regulating the dc voltage. It maintains the output voltage constant even through the current through it changes.

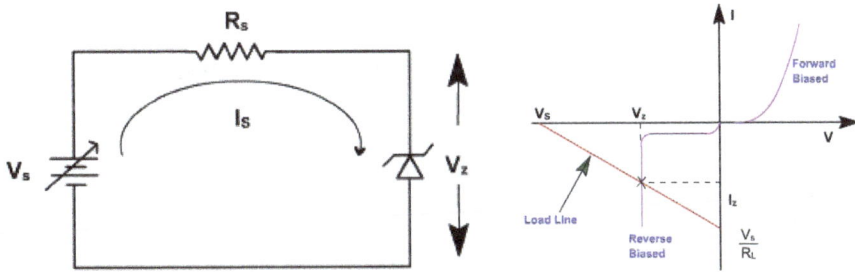

The load line of the circuit is given by $V_s = I_s R_s + V_z$. The load line is plotted along with zener characteristic in the right figure. The intersection point of the load line and the zener characteristic gives the output voltage and zener current.

To operate the zener in breakdown region V_s should always be greater then V_z. R_s is used to limit the current. If the V_s voltage changes, operating point also changes simultaneously but voltage across zener is almost constant. The first approximation of zener diode is a voltage source of V_z magnitude and second approximation includes the resistance also. The two approximate equivalent circuits are shown in the left figure.

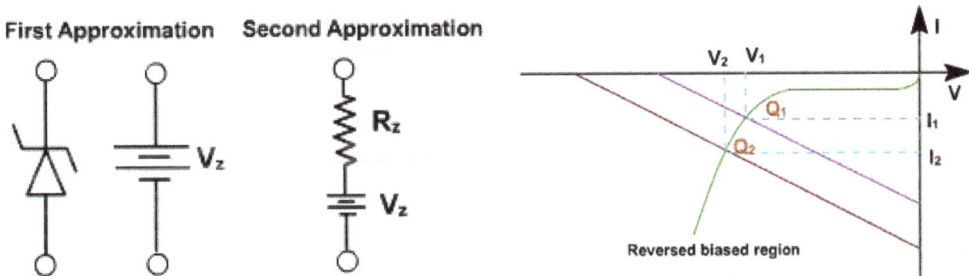

If second approximation of zener diode is considered, the output voltage varies slightly as shown in the right figure. The zener ON state resistance produces more I * R drop as the current increases. As the voltage varies form V_1 to V_2 the operating point shifts from Q_1 to Q_2.

The voltage at Q_1 is

$$V_1 = I_1 R_Z + V_Z$$

and at Q_2

$$V_2 = I_2 R_Z + V_Z$$

Thus, change in voltage is

$$V_2 - V_1 = (I_2 - I_1) R_Z$$

$$\Delta V_Z = \Delta I_Z R_Z$$

Design of Zener Regulator Circuit

A zenere regulator circuit is shown in the figure below. The varying load current is represented by a variable load resistance R_L.

The zener will work in the breakdown region only if the Thevenin voltage across zener is more than V_Z.

$$V_{TH} = v_s \frac{RL}{R_s + R_1}$$

If zener is operating in breakdown region, the current through R_s is given by

$$I_s = \frac{V_s - V_z}{R_s}$$

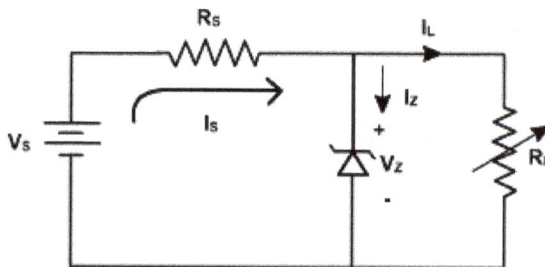

and load current $I_L = \dfrac{V_z}{R_L}$

$$I_s = I_z + I_L$$

The circuit is designed such that the diode always operates in the breakdown region and the voltage V_Z across it remains fairly constant even though the current I_Z through it vary considerably.

If the load I_L should increase, the current I_Z should decrease by the same percentage in order to maintain load current constant I_s. This keeps the voltage drop across R_s constant and hence the output voltage.

If the input voltage should increase, the zener diode passes a larger current, that extra voltage is dropped across the resistance R_s. If input voltage falls, the current I_Z falls such that V_Z is constant.

In the practical application the source voltage, v_s, varies and the load current also varies. The design challenge is to choose a value of R_s which permits the diode to maintain a relatively constant output voltage, even when the input source voltage varies and the load current also varies.

We now analyze the circuit to determine the proper choice of R_s. For the circuit shown in figure,

$$Rs = \frac{v_s - V_z}{i_R} = \frac{v_s - V_z}{i_z + i_L} \quad \text{(E-12)}$$

$$i_z = \frac{v_s - V_z}{R_s} - i_L \quad \text{(E-13)}$$

The variable quantities in Equation (E-13) are v_z and i_L. In order to assure that the diode remains in the constant voltage (breakdown) region, we examine the two extremes of input/output conditions, as follows:

- The current through the diode, i_z, is a minimum $(I_{Z\,min})$ when the load current, i_L is maximum $(I_{L\,max})$ and the source voltage, v_s is minimum $(V_{s\,min})$.

- The current through the diode, i_z, is a maximum $(I_{Z\,max})$ when the load current, i_L, is minmum $(i_{L\,min})$ and the source voltage v_s is minimum$(V_{s\,max})$.

When these characteristics of the two extremes are inserted into Equation (E-12),

we find $$R_s = \frac{V_{S_{min}} - V_z}{I_{L_{max}} + I_{z_{min}}} = \frac{V_{S_{max}} - V_z}{I_{L_{min}} + I_{z_{max}}} \quad \text{(E-14)}$$

$$\left(V_{S_{min}} - V_z\right)\left(I_{L_{min}} + I_{z_{max}}\right) = \left(V_{S_{max}} - V_z\right)\left(I_{L_{max}} + I_{z_{min}}\right) \quad \text{(E-15)}$$

In a practical problem, we know the range of input voltages, the range of output load currents, and the desired Zener voltage. Equation (E-15) thus represents one equation in two unknowns, the maximum and minimum Zener current. A second equation is found from the characteristic of zener. To avoid the non-constant portion of the characteristic curve, we use an accepted rule of thumb that the minimum Zener current should be 0.1 times the maximum (i.e., 10%), that is,

$$I_{z_{min}} = 0.1 \times I_{z_{max}} \quad \text{(E-16)}$$

Solving the equations E-15 and E-16, we get,

$$I_{z_{max}} = \frac{I_{L_{min}}\left(V_z - V_{S_{min}}\right) + I_{L_{max}}\left(V_{S_{max}} - V_z\right)}{V_{S_{min}} - 0.9V_z - 0.1V_{s_{max}}} \quad \text{(E-17)}$$

Now that we can solve for the maximum Zener current, the value of R_s, is calculated from Equation (E-14).

Zener diodes are manufactured with breakdown voltages V_z in the range of a few volts to a few hundred volts. The manufacturer specifies the maximum power the diode can dissipate. For example, a 1W, 10 V zener can operate safely at currents up to 100mA.

References

- Lidia Łukasiak & Andrzej Jakubowski (January 2010). "History of Semiconductors" (PDF). Journal of Telecommunication and Information Technology: 3

- Pugh, Emerson W.; Johnson, Lyle R.; Palmer, John H. (1991). IBM's 360 and early 370 systems. MIT Press. p. 34. ISBN 0-262-16123-0

- Guthrie, Frederick (October 1873) "On a relation between heat and static electricity," The London, Edinburgh and Dublin Philosophical Magazine and Journal of Science, 4th series, 46 : 257–266

- Shockley, William (1950). Electrons and holes in semiconductors : with applications to transistor electronics. R. E. Krieger Pub. Co. ISBN 0-88275-382-7

- Jonscher, A. K. (1961). "The physics of the tunnel diode". British Journal of Applied Physics. 12 (12): 654. Bibcode:1961BJAP...12..654J. doi:10.1088/0508-3443/12/12/304

- Louis Nashelsky, Robert L.Boylestad. Electronic Devices and Circuit Theory (9th ed.). India: Prentice-Hall of India Private Limited. pp. 7–10. ISBN 978-81-203-2967-6

- Lowe, Doug (2013). "Electronics Components: Diodes". Electronics All-In-One Desk Reference For Dummies. John Wiley & Sons. Retrieved January 4, 2013

- Crecraft, David; Stephen Gergely (2002). Analog Electronics: Circuits, Systems and Signal Processing. Butterworth-Heinemann. p. 110. ISBN 0-7506-5095-8

- Guarnieri, M. (2012). "The age of vacuum tubes: Early devices and the rise of radio communications". IEEE Ind. Electron. M. 6 (1): 41–43. doi:10.1109/MIE.2012.2182822

- Horowitz, Paul; Winfield Hill (1989). The Art of Electronics, 2nd Ed. London: Cambridge University Press. p. 44. ISBN 0-521-37095-7

- As in the Mott formula for conductivity, see Cutler, M.; Mott, N. (1969). "Observation of Anderson Localization in an Electron Gas". Physical Review. 181 (3): 1336. Bibcode:1969PhRv..181.1336C. doi:10.1103/PhysRev.181.1336

- Gupta, K. M.; Gupta, Nishu (2015). Advanced Semiconducting Materials and Devices. Springer. p. 236. ISBN 9783319197586

- J. W. Allen (1960). "Gallium Arsenide as a semi-insulator". Nature. 187 (4735): 403–405. Bibcode:1960Natur.187..403A. doi:10.1038/187403b0

- Andrei Grebennikov (2011). "§2.1.1: Diodes: Operational principle". RF and Microwave Transmitter Design. J Wiley & Sons. p. 59. ISBN 0-470-52099-X

- Elhami Khorasani, A.; Griswold, M.; Alford, T. L. (2014). "Gate-Controlled Reverse Recovery for Characterization of LDMOS Body Diode". IEEE Electron Device Letters. 35 (11): 1079. Bibcode:2014IEDL...35.1079E. doi:10.1109/LED.2014.2353301

- Riordan, Michael; Lillian Hoddeson (1988). Crystal fire: the invention of the transistor and the birth of the information age. USA: W. W. Norton & Company. pp. 88–97. ISBN 0-393-31851-6

- Redhead, P. A. (1998-05-01). "The birth of electronics: Thermionic emission and vacuum". Journal of Vacuum Science & Technology A: Vacuum, Surfaces, and Films. 16 (3): 1394. ISSN 0734-2101. doi:10.1116/1.581157

- Luque, Antonio; Steven Hegedus (29 March 2011). Handbook of Photovoltaic Science and Engineering. John Wiley & Sons. ISBN 978-0-470-97612-8

2

Electronic Circuits: An Integrated Study

An assimilation of various electronic components such as diodes, transistors, resistors, capacitors, and inductors that are connected with a copper wire form a complete circuit. It allows for the flow of electricity and is known as an electronic circuit. Apart from the common function of a diode to allow electricity to pass through one end and resist through other, they can be crafted to perform other special functions. By changing the semiconductor element as well as the impurity, special-purpose diodes could be manufactured. The following chapter elucidates the various techniques related to electronic circuits.

Electronic Circuit

The die from an Intel 8742, an 8-bit microcontroller that includes a CPU, 128 bytes of RAM, 2048 bytes of EPROM, and I/O "data" on current chip.

An electronic circuit is composed of individual electronic components, such as resistors, transistors, capacitors, inductors and diodes, connected by conductive wires or traces through which electric current can flow. The combination of components and wires allows various simple and complex operations to be performed: signals can be amplified, computations can be performed, and data can be moved from one place to another.

Circuits can be constructed of discrete components connected by individual pieces of wire, but today it is much more common to create interconnections by photolithographic techniques on a laminated substrate (a printed circuit board or PCB) and solder the components to these interconnections to create a finished circuit. In an integrated circuit or IC, the components and interconnections are formed on the same substrate, typically a semiconductor such as silicon or (less commonly) gallium arsenide.

A circuit built on a printed circuit board (PCB).

An electronic circuit can usually be categorized as an analog circuit, a digital circuit, or a mixed-signal circuit (a combination of analog circuits and digital circuits).

Breadboards, perfboards, and stripboards are common for testing new designs. They allow the designer to make quick changes to the circuit during development.

Analog Circuits

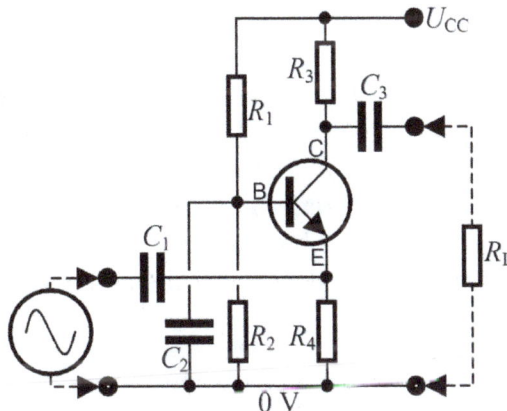

A circuit diagram representing an analog circuit, in this case a simple amplifier

Analog electronic circuits are those in which current or voltage may vary continuously with time to correspond to the information being represented. Analog circuitry is constructed from two fundamental building blocks: series and parallel circuits.

In a series circuit, the same current passes through a series of components. A string of Christmas lights is a good example of a series circuit: if one goes out, they all do. In a parallel circuit, all the components are connected to the same voltage, and the current divides between the various components according to their resistance.

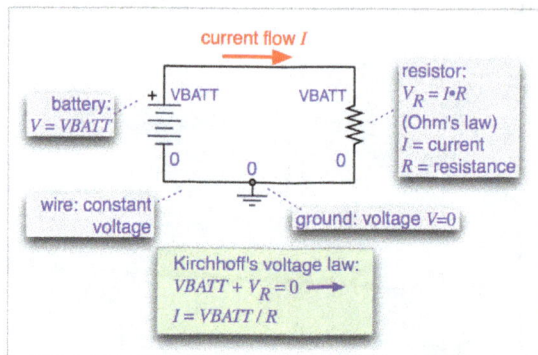

A simple schematic showing wires, a resistor, and a battery

The basic components of analog circuits are wires, resistors, capacitors, inductors, diodes, and transistors. (In 2012 it was demonstrated that memristors can be added to the list of available components.) Analog circuits are very commonly represented in schematic diagrams, in which wires are shown as lines, and each component has a unique symbol. Analog circuit analysis employs Kirchhoff's circuit laws: all the currents at a node (a place where wires meet), and the voltage around a closed loop of wires is 0. Wires are usually treated as ideal zero-voltage interconnections; any resistance or reactance is captured by explicitly adding a parasitic element, such as a discrete resistor or inductor. Active components such as transistors are often treated as controlled current or voltage sources: for example, a field-effect transistor can be modeled as a current source from the source to the drain, with the current controlled by the gate-source voltage.

When the circuit size is comparable to a wavelength of the relevant signal frequency, a more sophisticated approach must be used. Wires are treated as transmission lines, with (hopefully) constant characteristic impedance, and the impedances at the start and end determine transmitted and reflected waves on the line. Such considerations typically become important for circuit boards at frequencies above a GHz; integrated circuits are smaller and can be treated as lumped elements for frequencies less than 10GHz or so.

An alternative model is to take independent power sources and induction as basic electronic units; this allows modeling frequency dependent negative resistors, gyrators, negative impedance converters, and dependent sources as secondary electronic components.

Digital Circuits

In digital electronic circuits, electric signals take on discrete values, to represent logical and numeric values. These values represent the information that is being processed.

In the vast majority of cases, binary encoding is used: one voltage (typically the more positive value) represents a binary '1' and another voltage (usually a value near the ground potential, 0 V) represents a binary '0'. Digital circuits make extensive use of transistors, interconnected to create logic gates that provide the functions of Boolean logic: AND, NAND, OR, NOR, XOR and all possible combinations thereof. Transistors interconnected so as to provide positive feedback are used as latches and flip flops, circuits that have two or more metastable states, and remain in one of these states until changed by an external input. Digital circuits therefore can provide both logic and memory, enabling them to perform arbitrary computational functions. (Memory based on flip-flops is known as static random-access memory (SRAM). Memory based on the storage of charge in a capacitor, dynamic random-access memory (DRAM) is also widely used.)

The design process for digital circuits is fundamentally different from the process for analog circuits. Each logic gate regenerates the binary signal, so the designer need not account for distortion, gain control, offset voltages, and other concerns faced in an analog design. As a consequence, extremely complex digital circuits, with billions of logic elements integrated on a single silicon chip, can be fabricated at low cost. Such digital integrated circuits are ubiquitous in modern electronic devices, such as calculators, mobile phone handsets, and computers. As digital circuits become more complex, issues of time delay, logic races, power dissipation, non-ideal switching, on-chip and inter-chip loading, and leakage currents, become limitations to the density, speed and performance.

Digital circuitry is used to create general purpose computing chips, such as microprocessors, and custom-designed logic circuits, known as application-specific integrated circuit (ASICs). Field-programmable gate arrays (FPGAs), chips with logic circuitry whose configuration can be modified after fabrication, are also widely used in prototyping and development.

Mixed-signal Circuits

Mixed-signal or hybrid circuits contain elements of both analog and digital circuits. Examples include comparators, timers, phase-locked loops, analog-to-digital converters, and digital-to-analog converters. Most modern radio and communications circuitry uses mixed signal circuits. For example, in a receiver, analog circuitry is used to amplify and frequency-convert signals so that they reach a suitable state to be converted into digital values, after which further signal processing can be performed in the digital domain.

Bridge Circuit

A bridge circuit is a topology of electrical circuit in which two circuit branches (usually in parallel with each other) are "bridged" by a third branch connected between the first two branches at some intermediate point along them. The bridge was originally developed for laboratory measurement purposes and one of the intermediate bridging points is often adjustable when so used. Bridge circuits now find many applications, both linear and non-linear, including in instrumentation, filtering and power conversion.

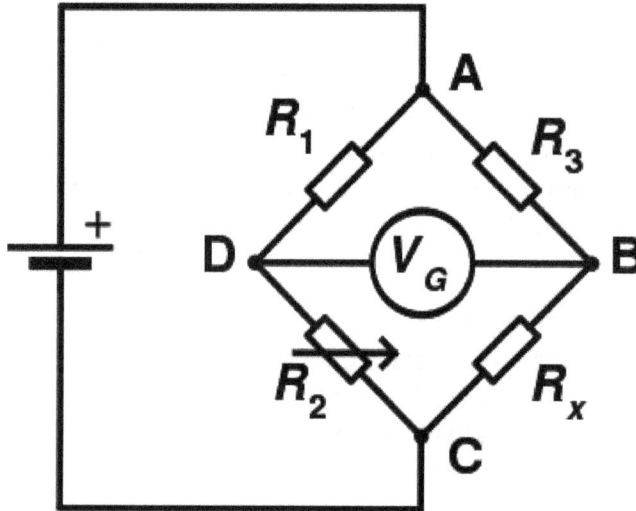

Schematic of a Wheatstone bridge

The best-known bridge circuit, the Wheatstone bridge, was invented by Samuel Hunter Christie and popularized by Charles Wheatstone, and is used for measuring resistance. It is constructed from four resistors, two of known values R_1 and R_3, one whose resistance is to be determined R_x, and one which is variable and calibrated R_2. Two opposite vertices are connected to a source of electric current, such as a battery, and a galvanometer is connected across the other two vertices. The variable resistor is adjusted until the galvanometer reads zero. It is then known that the ratio between the variable resistor and its neighbour R_1 is equal to the ratio between the unknown resistor and its neighbour R_3, which enables the value of the unknown resistor to be calculated.

The Wheatstone bridge has also been generalised to measure impedance in AC circuits, and to measure resistance, inductance, capacitance, and dissipation factor separately. Various arrangements are known as the Wien bridge, Maxwell bridge and Heaviside bridge. All are based on the same principle, which is to compare the output of two potentiometers sharing a common source.

In power supply design, a bridge circuit or bridge rectifier is an arrangement of diodes or similar devices used to rectify an electric current, i.e. to convert it from an unknown or alternating polarity to a direct current of known polarity.

In some motor controllers, an H-bridge is used to control the direction the motor turns.

Various types of bridge circuit:

Wheatstone bridge	Wien bridge	Maxwell bridge
H bridge	Kelvin bridge R7 represents parasitic resistance which can affect the accuracy.	Diode bridge
	/	
Fontana bridge	Lattice bridge	Bridged T circuit	Carey Foster bridge

Bridge Rectifier

The single – phase full wave bridge rectifier is shown in the figure below. It is the most widely used rectifier. It also provides currents in both the half cycle of input supply.

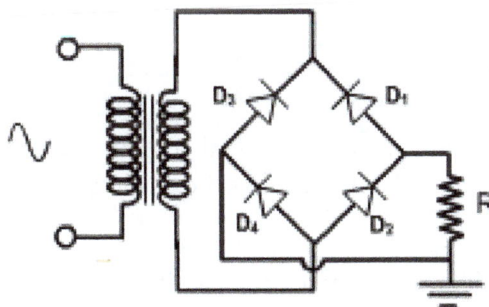

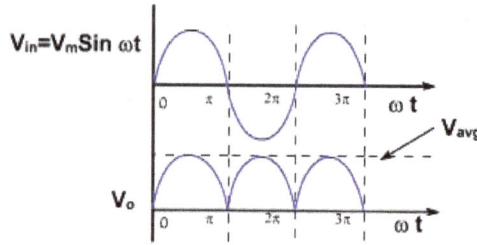

In the positive half cycle, D_1 & D_4 are forward biased and D_2 & D_3 are reverse biased. In the negative half cycle, D_2 & D_3 are forward biased, and D_1 & D_4 are reverse biased. The output voltage waveform is shown in the above figure and it is same as full wave rectifier but the advantage is that PIV rating of diodes are V m and only single secondary transformer is required.

The main disadvantage is that it requires four diodes. When low dc voltage is required then secondary voltage is low and diodes drop (1.4V) becomes significant. For low dc output, 2-pulse center tap rectifier is used because only one diode drop is there.

The ripple factor is the measure of the purity of dc output of a rectifier and is defined as

$$Ripple\ factor = \frac{r.m.s\ value\ of\ the\ ac\ output\ voltage}{average\ dc\ output\ voltage}$$

$$= \sqrt{V_0^2 + \sum_{n=1}^{\infty} V_n^2}$$

Therefore,

$$Ripple\ factor = \frac{\sqrt{V_{rms}^2 - V_0^2}}{V_0}$$

$$= \sqrt{\left(\frac{V_{rms}}{V_0}\right)^2 - 1}$$

Clipper (Electronics)

In electronics, a clipper is a device designed to prevent the output of a circuit from exceeding a predetermined voltage level without distorting the remaining part of the applied waveform.

A clipping circuit consists of linear elements like resistors and non-linear elements like junction diodes or transistors, but it does not contain energy-storage elements like capacitors. Clipping circuits are used to select for purposes of transmission, that part of a signal wave form which lies above or below a certain reference voltage level.

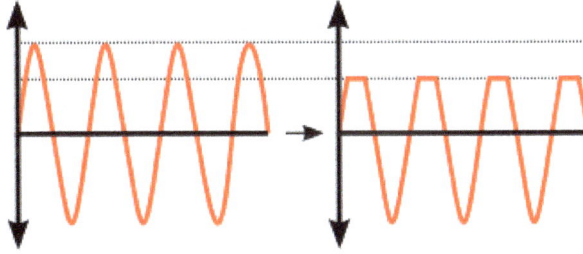

Voltage clipping limits the voltage to a device without affecting the rest of the waveform

Thus a clipper circuit can remove certain portions of an arbitrary waveform near the positive or negative peaks. Clipping may be achieved either at one level or two levels. Usually under the section of clipping, there is a change brought about in the wave shape of the signal.

Clipping circuits are also called slicers or amplitude selectors.

Types

Diode Clipper

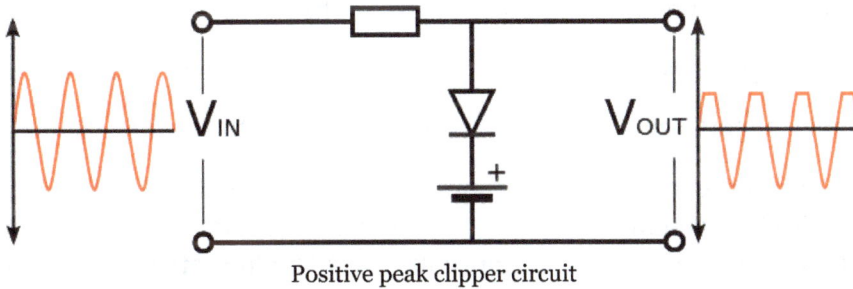

Positive peak clipper circuit

A simple diode clipper can be made with a diode and a resistor. This will remove either the positive, or the negative half of the waveform depending on the direction the diode is connected. The simple circuit clips at zero voltage (or to be more precise, at the small forward voltage of the forward biased diode) but the clipping voltage can be set to any desired value with the addition of a reference voltage. The diagram illustrates a positive reference voltage but the refcrence can be positive or negative for both positive and negative clipping giving four possible configurations in all.

The simplest circuit for the voltage reference is a resistor potential divider connected between the voltage rails. This can be improved by replacing the lower resistor with a zener diode with a breakdown voltage equal to the required reference voltage. The zener acts as a voltage regulator stabilising the reference voltage against supply and load variations.

Zener Diode

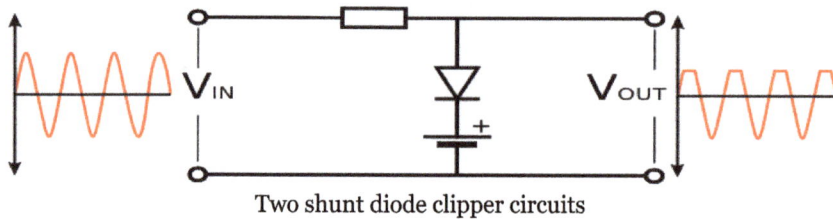

Two shunt diode clipper circuits

In the example circuit on the right, two zener diodes are used to clip the voltage V_{IN}. The voltage in either direction is limited to the reverse breakdown voltage *plus* the voltage drop across one zener diode.

Op-amp Precision Clipper

For very small values of clipping voltage on low-level signals the I-V curve of the diode can result in clipping onset that is not very sharp. Precision clippers can be made by placing the clipping device in the feedback circuit of an operational amplifier in a similar way to precision rectifiers.

Classification

Clippers may be classified into two types based on the positioning of the diode.

- Series Clippers, where the diode is in series with the load resistance, and
- Shunt Clippers, where the diode is shunted across the load resistance.

The diode capacitance affects the operation of the clipper at high frequency and influences the choice between the above two types. High frequency signals are attenuated in the shunt clipper as the diode capacitance provides an alternative path to output current. In the series clipper, clipping effectiveness is reduced for the same reason as the high frequency current passes through without being sufficiently blocked.

Clippers may be classified based on the orientation(s) of the diode. The orientation decides which half cycle is affected by the clipping action.

The clipping action can be made to happen at an arbitrary level by using a biasing elements (potential sources) in series with the diode.

- Positively Biased Diode Clipper
- Negatively Biased Diode Clipper

The signal can be clipped to between two levels by using both types of diode clippers in combination. This clipper is referred to as

- Combinational Diode Clipper or Two-Level Clippers

The clamping network is the one that will "clamp" a signal to a different dc level. The network must have capacitor, a diode, and a resistive element, but it also employs an independent dc supply to introduce an additional shift.

Clippers

Clipping circuits are used to select that portion of the input wave which lies above or below some reference level. Some of the clipper circuits are discussed here. The transfer characteristic (v_o vs v_i) and the output voltage waveform for a given input voltage are also discussed.

Circuit 1

The circuit shown in the first figure, clips the input signal above a reference voltage (V_R).

In this clipper circuit,

If $v_i < V_R$, diode is reversed biased and does not conduct. Therefore, $v_o = v_i$

and, if $v_i > V_R$, diode is forward biased and thus, $v_o = V_R$.

The transfer characteristic of the clippers is shown in the second figure.

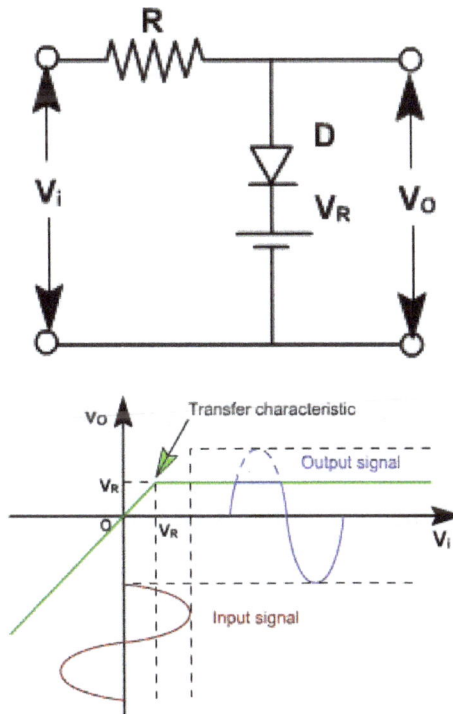

Circuit 2

The clipper circuit shown in the first figure clips the input signal below reference voltage V_R.

In this clipper circuit,

If $v_i > V_R$, diode is reverse biased. $v_o = v_i$

and,   If $v_i < V_R$, diode is forward biased. $v_o = V_R$

The transfer characteristic of the circuit is shown in the second figure.

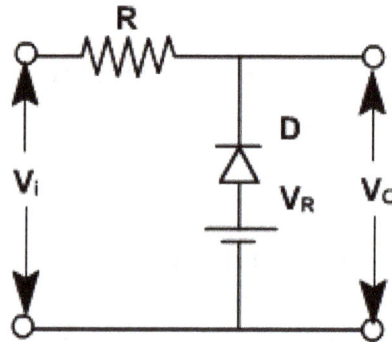

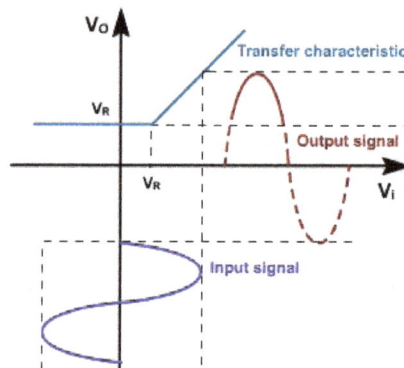

Circuit 3

To clip the input signal between two independent levels ($V_{R1} < V_{R2}$), the clipper circuit is shown in the left figure.

The diodes D_1 & D_2 are assumed ideal diodes.

For this clipper circuit, when $v_i \leq V_{R1}$, $v_o = V_{R1}$

and, $v_i \geq V_{R2}$, $v_o = V_{R2}$

and, $V_{R1} < v_i < V_{R2}$ $v_o = v_i$

The transfer characteristic of the clipper is shown in the right figure.

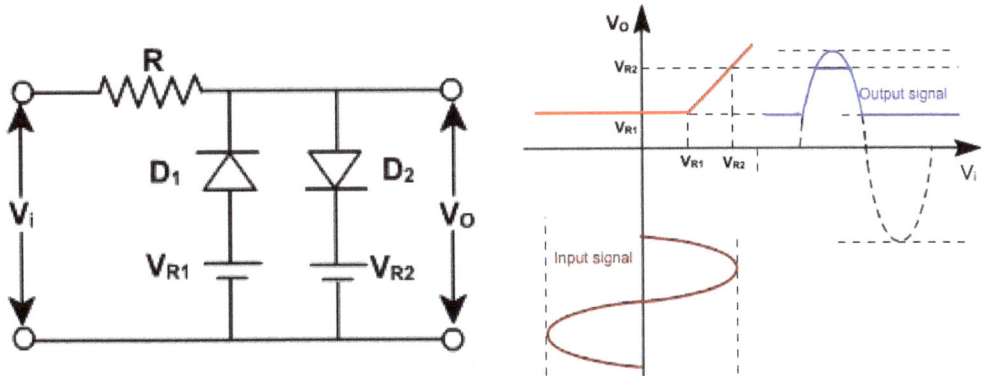

Example – 1

Draw the transfer characteristic of the circuit shown in the figure.

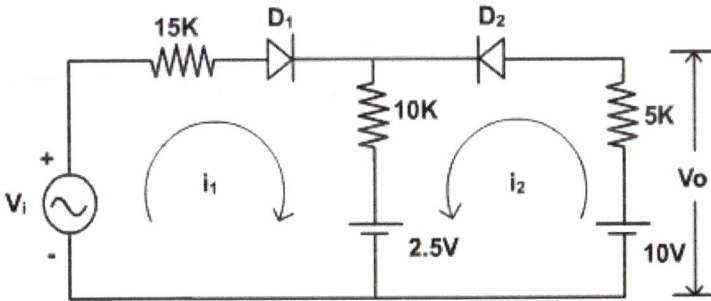

Solution

When diode D_1 is off, $i_1 = 0$, D_2 must be ON.

$$i_2 = \frac{10 - 2.5}{10k + 5k} = \frac{7.5}{15k} = 0.5mA$$

and $v_o = 10 - 5 \times 0.25 = 7.5$ V

$v_p = v_o = 7.5$ V

Therefore, D_1 is reverse biased only if $v_i < 7.5$ V

If D_2 is off and D_1 is ON, $i_2 = 0$

$$i_1 = \frac{v_1 - 2.5}{15k + 10k} = 0.04v_i - 0.1$$

and $\quad v_p = 10\,(\,0.04\ v_i - 0.1\,) + 2.5 = 0.4\ v_i + 1.5$

For D_2 to be reverse biased,

$$v_p > 10V$$
$$0.04v_i + 1.5 > 10V$$
$$0.4v_i > 8.5V$$
$$v_i > 21.25V$$

Between 7.5 V and 21.25 V both the diodes are ON.

$$v_i = 15i_1 + 10(i_1 + i_2) + 2.5$$
$$10 = 5i_2 + 10(i_1 + i_2) + 2.5$$
$$i_2 = \frac{10v_i - 250 - 25 + 62.5}{100 - 375}$$
$$v_0 = 10 - 5i_2$$
$$= 10 - 5\left(\frac{10v_i - 275 + 62.5}{-275}\right)$$
$$= 10 + \frac{50v_i - 1375 + 312.5}{275}$$
$$v_0 = \frac{2v_i + 67.5}{11}$$

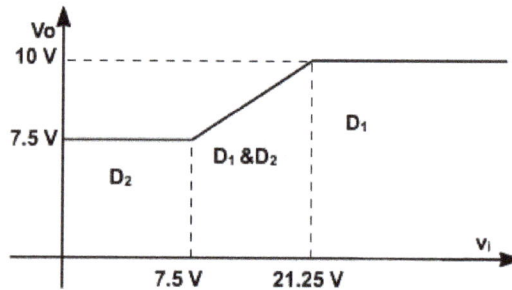

The transfer characteristic of the circuit is shown in the above figure.

Clipper and Clamper Circuits

Clippers

In the clipper circuits, discussed so far, diodes are assumed to be ideal device. If third approximation circuit of diode is used, the transfer characteristics of the clipper circuits will be modified.

Circuit 4

Consider the clipper circuit shown in the left figure to clip the input signal above reference voltage

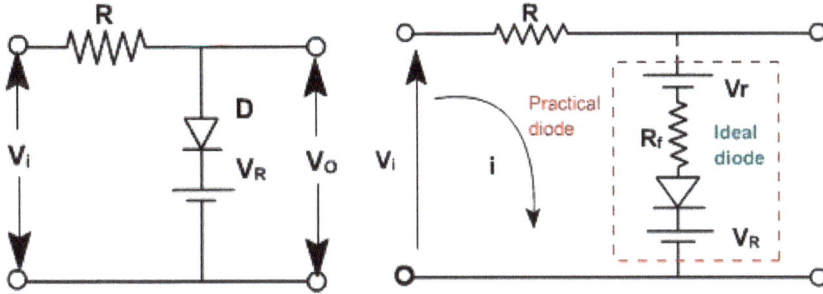

When $v_i < (V_R + V_r)$, diode D is reverse biased and therefore, $v_o = v_i$.

and when $v_i > (V_R + V_r)$, diode D is forward biased and conducts. The equivalent circuit, in this case is shown in figure.

The current i in the circuit is given by

$$i = \frac{v_i - V_r - V_R}{R + R_f}$$

$$v_0 = V_r + V_R + iR_f$$

$$= V_r + V_R + \frac{(V_i - V_r - V_R)R_f}{R + R_f}$$

$$v_0 = (V_r + V_R)\frac{R}{R + R_f} + \frac{R_f}{R + R_f}v_i$$

The transfer characteristic of the circuit is shown in the figure below.

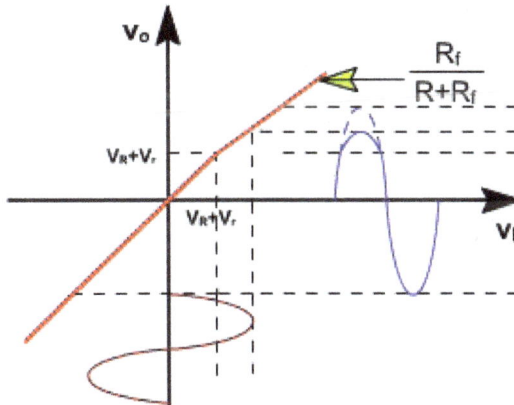

Circuit 5

Consider the clipper circuit shown in the figure on the left, which clips the input signal below the reference level (V_R).

If $v_i > (V_R - V_r)$, diode D is reverse biased, thus $v_0 = v_i$ and when $v_i < (v_R - V_r)$, D condcuts and the equivalent circuit becomes as shown in the figure on the right.

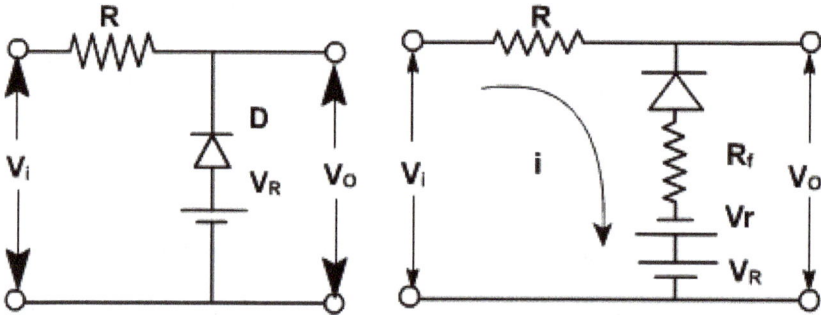

Therefore,

$$i = \frac{v_i + V_r + V_R}{R + R_f}$$

$$v_0 = iR_f + V_R - V_r$$

$$= \frac{R_f}{R + R_f}(V_i - (V_R - V_r)) + (V_R + V_r)$$

$$= \frac{R}{R + R_f}(V_R - V_r) + \frac{R_f V_i}{R + R_f}$$

The transfer characteristic of the circuit is shown in the figure below.

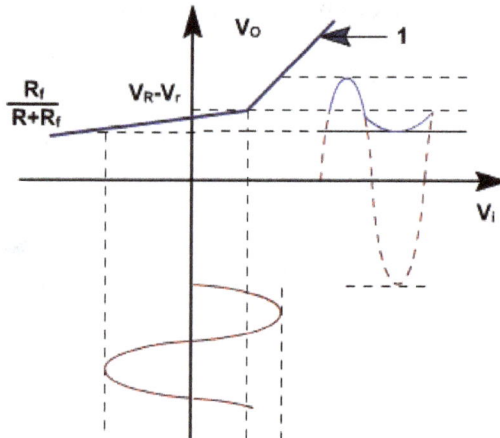

Example - 1

Find the output voltage v out of the clipper circuit of fig. (a) assuming that the diodes are

a.   ideal.

b. V_{on} = 0.7 V. For both cases, assume R_F is zero.

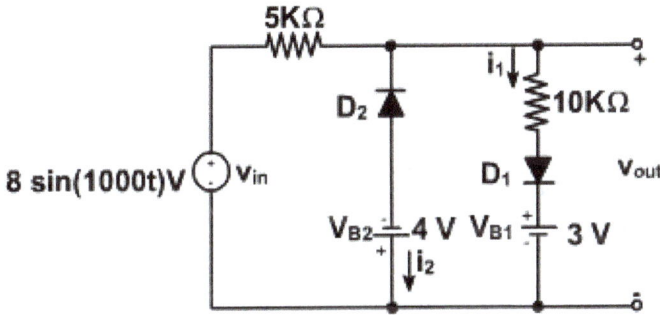

Fig. (a)

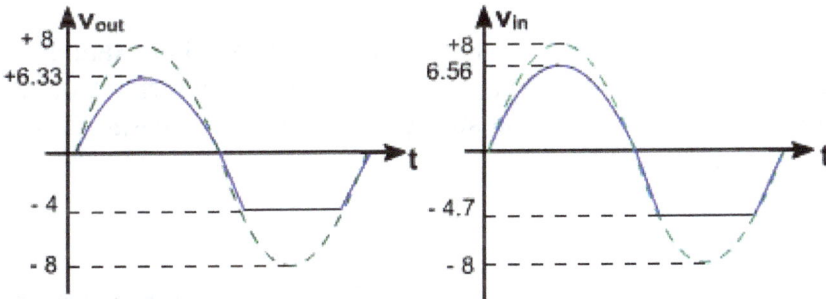

Fig. (b)

Solution

(a). When v_{in} is positive and v_{in} < 3, then v_{out} = v_{in}

and when v_{in} is positive and v_{in} > 3, then

$$i_1 = \frac{v_{in} - 3}{1.5 \times 10^4}$$

$$v_{out} = 10^4 i_1 + 3 = \frac{2}{3} v_{in} + 1$$

At v_{in} = 8 V(peak), v_{out} = 6.33 V.

When v_{in} is negative and v_{in} > - 4, then v_{out} = v_{in}

When v_{in} is negative and v_{in} < -4, then v_{out} = -4V

The resulting output wave shape is shown in fig. (b).

(b). When V_{ON} = 0.7 V, v_{in} is positive and v_{in} < 3.7 V, then v_{out} = v_{in}

When v_{in} > 3.7 V, then

$$i_1 = \frac{v_{in} - 3.7}{1.5 \times 10^4}$$

$$v_{out} = 10^4 i_1 + 3.7 = \frac{2}{3} v_{in} + 1.23$$

When v_{in} = 8V, v_{out} = 6.56 V.

When v_{in} is negative and v_{in} > -4.7 V, then v_{out} = v_{in}

When v_{in} < - 4.7 V, then v_{out} = - 4.7 V

The resulting output wave form is shown in fig. (b).

Clamper Circuits

Clamping is a process of introducing a dc level into a signal. For example, if the input voltage swings from -10 V and +10 V, a positive dc clamper, which introduces +10 V in the input will produce the output that swings ideally from 0 V to +20 V. The complete waveform is lifted up by +10 V.

Negative Diode Clamper

A negative diode clamper is shown in the left figure, which introduces a negative dc voltage equal to peak value of input in the input signal.

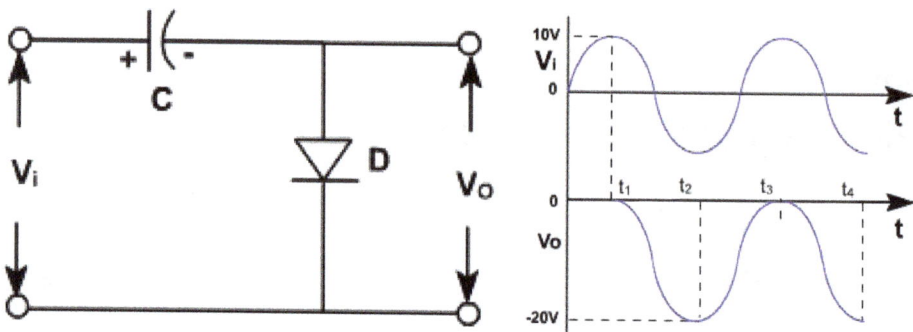

Let the input signal swings form +10 V to -10 V. During first positive half cycle as V i rises from 0 to 10 V, the diode conducts. Assuming an ideal diode, its voltage, which is also the output must be zero during the time from 0 to t_1. The capacitor charges during this period to 10 V, with the polarity shown.

At that V_i starts to drop which means the anode of D is negative relative to cathode, ($V_D = v_i - v_c$) thus reverse biasing the diode and preventing the capacitor from discharging shown in the figure on the right. Since the capacitor is holding its charge it behaves as a DC voltage source while the diode appears as an open circuit, therefore the equivalent circuit becomes an input supply in series with -10 V dc voltage as shown in the figure below, and the resultant output voltage is the sum of instantaneous input voltage and dc voltage (-10 V).

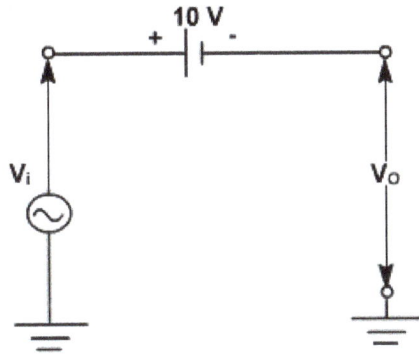

Clamper (Electronics)

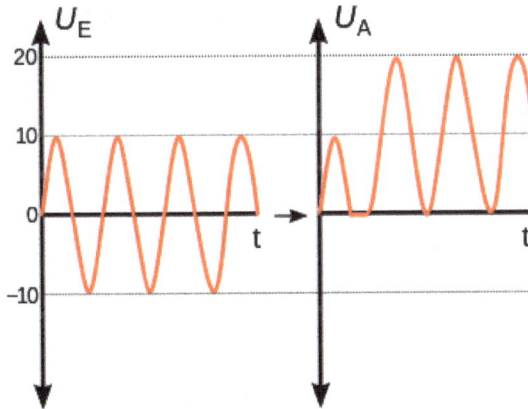

Positive unbiased voltage clamping vertically translates the input waveform so that all parts of it are approximately greater than 0 V. Note that the negative swing of the output will not dip below about −0.6 V, assuming a silicon pn diode.

A clamper is an electronic circuit that fixes either the positive or the negative peak excursions of a signal to a defined value by shifting its DC value. The clamper does not restrict the peak-to-peak excursion of the signal, it moves the whole signal up or down so as to place the peaks at the reference level. A diode clamp (a simple, common type) consists of a diode, which conducts electric current in only one direction and prevents the signal exceeding the reference value; and a capacitor which provides a DC offset from the stored charge. The capacitor forms a time constant with the resistor load which determines the range of frequencies over which the clamper will be effective.

General Function

A clamping circuit (also known as a clamper) will bind the upper or lower extreme of a waveform to a fixed DC voltage level. These circuits are also known as DC voltage restorers. Clampers can be constructed in both positive and negative polarities. When unbiased, clamping circuits will fix the voltage lower limit (or upper limit, in the case of negative clampers) to 0 Volts. These circuits clamp a peak of a waveform to a specific DC level compared with a capacitively coupled signal which swings about its average DC level.

Types

Clamp circuits are categorised by their operation; negative or positive, and biased or unbiased. A positive clamp circuit(negative peak clamper) outputs a purely positive waveform from an input signal; it offsets the input signal so that all of the waveform is greater than 0 V. A negative clamp is the opposite of this - this clamp outputs a purely negative waveform from an input signal. A bias voltage between the diode and ground offsets the output voltage by that amount.

For example, an input signal of peak value 5 V (V_{INpeak} = 5 V) is applied to a positive clamp with a bias of 3 V (V_{BIAS} = 3 V), the peak output voltage will be:

$V_{OUTpeak} = 2 * V_{INpeak} + V_{BIAS}$

$V_{OUTpeak} = 2 * 5 V + 3 V$

$V_{OUTpeak} = 13 V$

Positive Unbiased

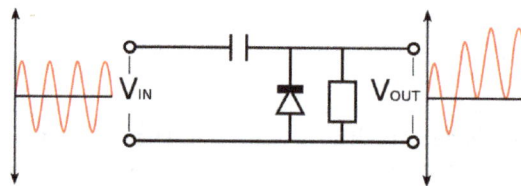

A positive unbiased clamp.

In the negative cycle of the input AC signal, the diode is forward biased and conducts, charging the capacitor to the peak negative value of V_{IN}. During the positive cycle, the diode is reverse biased and thus does not conduct. The output voltage is therefore equal to the voltage stored in the capacitor plus the input voltage, so $V_{OUT} = V_{IN} + V_{INpeak}$. This is also called a Villard circuit.

Negative Unbiased

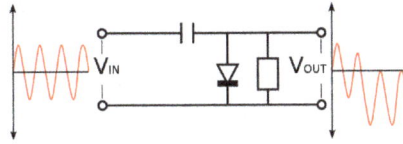

A negative unbiased clamp

A negative unbiased clamp is the opposite of the equivalent positive clamp. In the positive cycle of the input AC signal, the diode is forward biased and conducts, charging the capacitor to the peak positive value of V_{IN}. During the negative cycle, the diode is reverse biased and thus does not conduct. The output voltage is therefore equal to the voltage stored in the capacitor plus the input voltage again, so $V_{OUT} = V_{IN} - V_{INpeak}$

Positive Biased

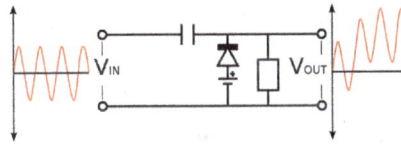

A positive biased clamp

A positive biased voltage clamp is identical to an equivalent unbiased clamp but with the output voltage offset by the bias amount V_{BIAS}. Thus, $V_{OUT} = V_{IN} + (V_{INpeak} + V_{BIAS})$

Negative Biased

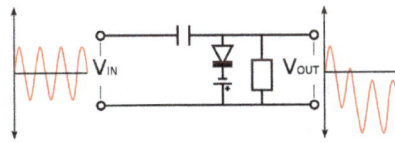

A negative biased clamp

A negative biased voltage clamp is likewise identical to an equivalent unbiased clamp but with the output voltage offset in the negative direction by the bias amount V_{BIAS}. Thus, $V_{OUT} = V_{IN} - (V_{INpeak} + V_{BIAS})$

Op-amp Circuit

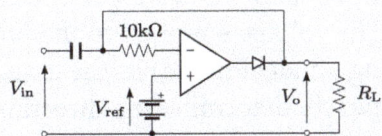

Precision op-amp clamp circuit

The figure shows an op-amp clamp circuit with a non-zero reference clamping voltage. The advantage here is that the clamping level is at precisely the reference voltage. There is no need to take into account the forward volt drop of the diode (which is necessary in the preceding simple circuits as this adds to the reference voltage). The effect of the diode volt drop on the circuit output will be divided down by the gain of the amplifier, resulting in an insignificant error. The circuit also has a great improvement in linearity at small input signals in comparison to the simple diode circuit.

Clamping for Input Protection

Clamping can be used to adapt an input signal to a device that cannot make use of or may be damaged by the signal range of the original input.

Principles of Operation

In general the clamper includes a capacitor, followed by a diode in parallel with the load. The clamper circuit relies on a change in the capacitor's time constant which is the result of the diode changing the current path of the capacitor with the changing polarity of the AC input voltage. The magnitude of R and C are chosen so that the time constant, $\tau = RC$, is large enough to ensure that the voltage across the capacitor does not discharge significantly during the diode's non-conducting interval. However this kind of discharge only happens when the load resistor is very big, which takes the capacitor a lot of time to discharge, and can be ignored at high frequencies. On the other hand, the capacitor is chosen small enough to allow it to charge quickly during the diode's conducting interval. At the same time, the AC supply voltage frequency should be chosen low enough for the capacitor to fully charge in one quarter of a cycle.

During the first negative phase of the AC input voltage, the capacitor in a positive clamper circuit charges rapidly. As V_{in} becomes positive, the capacitor serves as a voltage doubler; since it has stored the equivalent of V_{in} during the negative cycle, it provides nearly that voltage during the positive cycle. This essentially doubles the voltage seen by the load. As V_{in} becomes negative, the capacitor acts as a battery of the same voltage of V_{in}. The voltage source and the capacitor counteract each other, resulting in a net voltage of zero as seen by the load.

Biased Versus Non-biased

By using a voltage source and resistor, the clamper can be biased to bind the output voltage to a different value. The voltage supplied to the potentiometer will be equal to the offset from zero (assuming an ideal diode) in the case of either a positive or negative clamper (the clamper type will determine the direction of the offset). If a negative voltage is supplied to either positive or negative, the waveform will cross the x-axis and be bound to a value of this magnitude on the opposite side. Zener diodes can also be used in place of a voltage source and potentiometer, hence setting the offset at the Zener voltage.

Examples

Clamping circuits were common in analog television receivers. These sets have a DC restorer circuit, which returns the voltage of the video signal during the 'back porch' of the line blanking (retrace) period to 0 V. Low frequency interference, especially power line hum, induced onto the signal spoils the rendering of the image, and in extreme cases causes the set to lose synchronization. This interference can be effectively removed via this method.

Positive Clamper:

The positive clamper circuit is shown in the left figure, which introduces positive dc voltage equal to the peak of input signal. The operation of the circuit is same as of negative clamper.

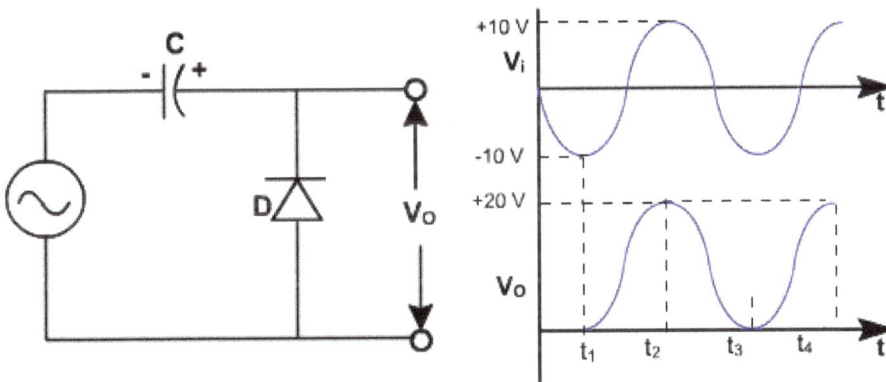

Let the input signal swings form +10 V to -10 V. During first negative half cycle as V_i rises from 0 to -10 V, the diode conducts. Assuming an ideal diode, its voltage, which is also the output must be zero during the time from 0 to t_1. The capacitor charges during this period to 10 V, with the polarity shown.

After that V_i starts to drop which means the anode of D is negative relative to cathode, $(V_D = v_i - v_c)$ thus reverse biasing the diode and preventing the capacitor from discharging in the right figure. Since the capacitor is holding its charge it behaves as a DC voltage source while the diode appears as an open circuit, therefore the equivalent circuit becomes an input supply in series with +10 V dc voltage and the resultant output voltage is the sum of instantaneous input voltage and dc voltage (+10 V).

To clamp the input signal by a voltage other than peak value, a dc source is required. As shown in figure, the dc source is reverse biasing the diode.

The input voltage swings from +10 V to -10 V. In the negative half cycle when the voltage exceed 5V then D conduct. During input voltage variation from −5 V to -10 V, the capacitor charges to 5 V with the polarity shown in the left figure below. After that D becomes reverse biased and open circuited. Then complete ac signal is shifted upward by 5 V. The output waveform is shown in the right figure below.

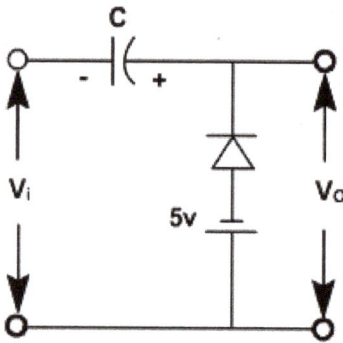

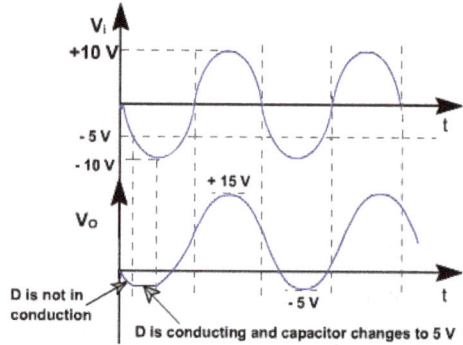

Voltage Doubler

A voltage doubler is an electronic circuit which charges capacitors from the input voltage and switches these charges in such a way that, in the ideal case, exactly twice the voltage is produced at the output as at its input.

The simplest of these circuits are a form of rectifier which take an AC voltage as input and outputs a doubled DC voltage. The switching elements are simple diodes and they are driven to switch state merely by the alternating voltage of the input. DC-to-DC voltage doublers cannot switch in this way and require a driving circuit to control the switching. They frequently also require a switching element that can be controlled directly, such as a transistor, rather than relying on the voltage across the switch as in the simple AC-to-DC case.

Voltage doublers are a variety of voltage multiplier circuit. Many, but not all, voltage doubler circuits can be viewed as a single stage of a higher order multiplier: cascading identical stages together achieves a greater voltage multiplication.

Voltage Doubling Rectifiers

Villard Circuit

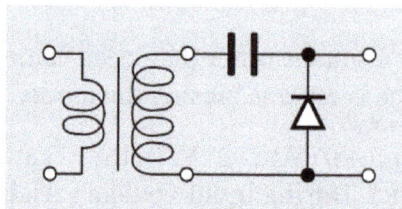

Villard circuit

The Villard circuit, due to Paul Ulrich Villard, consists simply of a capacitor and a diode. While it has the great benefit of simplicity, its output has very poor ripple characteristics. Essentially, the circuit is a diode clamp circuit. The capacitor is charged on the negative half cycles to the peak AC voltage (V_{pk}). The output is the superposition of the input AC waveform and the steady DC of the capacitor. The effect of the circuit is to shift the DC value of the waveform. The negative peaks of the AC waveform are "clamped" to 0 V (actually $-V_F$, the small forward bias voltage of the diode) by the diode, therefore the positive peaks of the output waveform are $2V_{pk}$. The peak-to-peak ripple is an enormous $2V_{pk}$ and cannot be smoothed unless the circuit is effectively turned into one of the more sophisticated forms. This is the circuit (with diode reversed) used to supply the negative high voltage for the magnetron in a microwave oven.

Greinacher Circuit

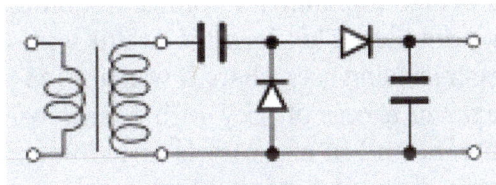

Greinacher circuit

The Greinacher voltage doubler is a significant improvement over the Villard circuit for a small cost in additional components. The ripple is much reduced, nominally zero under open-circuit load conditions, but when current is being drawn depends on the resistance of the load and the value of the capacitors used. The circuit works by following a Villard cell stage with what is in essence a peak detector or envelope detector stage. The peak detector cell has the effect of removing most of the ripple while preserving the peak voltage at the output. The Greinacher circuit is also commonly known as the half-wave voltage doubler.

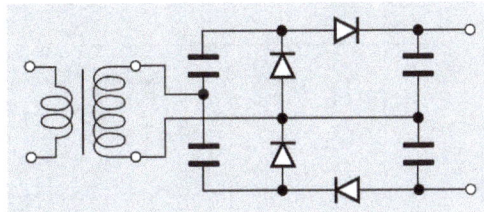

Voltage quadrupler – two Greinacher cells of opposite polarities

This circuit was first invented by Heinrich Greinacher in 1913 (published 1914) to provide the 200–300 V he needed for his newly invented ionometer, the 110 V AC supplied by the Zurich power stations of the time being insufficient. He later extended this idea into a cascade of multipliers in 1920. This cascade of Greinacher cells is often inaccurately referred to as a Villard cascade. It is also called a Cockcroft–Walton multiplier after the particle accelerator machine built by John Cockcroft and Ernest Walton, who independently discovered the circuit in 1932. The concept in this topology can be extended to a voltage quadrupler circuit by using two Greinacher cells of opposite polarities driven from the same AC source. The output is taken across the two individual outputs. As with a bridge circuit, it is impossible to simultaneously ground the input and output of this circuit.

Bridge Circuit

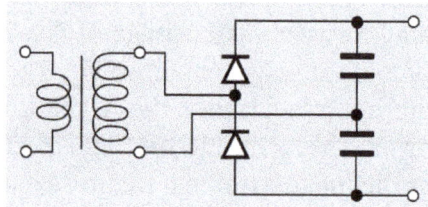

Bridge (Delon) voltage doubler

The Delon circuit uses a bridge topology for voltage doubling; consequently it is also called a full-wave voltage doubler. This form of circuit was, at one time, commonly found in cathode ray tube television sets where it was used to provide an e.h.t. voltage supply. Generating voltages in excess of 5 kV with a transformer has safety issues in terms of domestic equipment and in any case is uneconomical. However, black and white television sets required an e.h.t. of 10 kV and colour sets even more. Voltage doublers were used to either double the voltage on an e.h.t winding on the mains transformer or were applied to the waveform on the line flyback coils.

The circuit consists of two half-wave peak detectors, functioning in exactly the same way as the peak detector cell in the Greinacher circuit. Each of the two peak detector cells operates on opposite half-cycles of the incoming waveform. Since their outputs are in series, the output is twice the peak input voltage.

Switched Capacitor Circuits

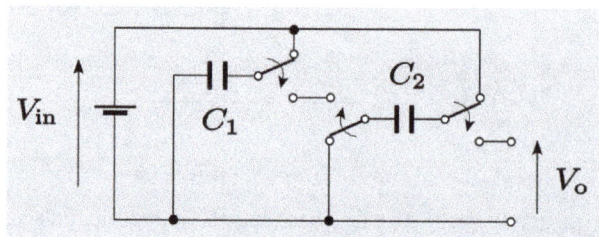

Switched capacitor voltage doubler achieved by simply switching charged capacitors from parallel to series

It is possible to use the simple diode-capacitor circuits described above to double the voltage of a DC source by preceding the voltage doubler with a chopper circuit. In effect, this converts the DC to AC before application to the voltage doubler. More efficient circuits can be built by driving the switching devices from an external clock so that both functions, the chopping and multiplying, are achieved simultaneously. Such circuits are known as switched capacitor circuits. This approach is especially useful in low-voltage battery-powered applications where integrated circuits require a voltage supply greater than the battery can deliver. Frequently, a clock signal is readily available on board the integrated circuit and little or no additional circuitry is needed to generate it.

Conceptually, perhaps the simplest switched capacitor configuration is that shown schematically in the figure above. Here two capacitors are simultaneously charged to the same voltage in parallel. The supply is then switched off and the capacitors are switched into series. The output is taken from across the two capacitors in series resulting in an output double the supply voltage. There are many different switching devices that could be used in such a circuit, but in integrated circuits MOSFET devices are frequently employed.

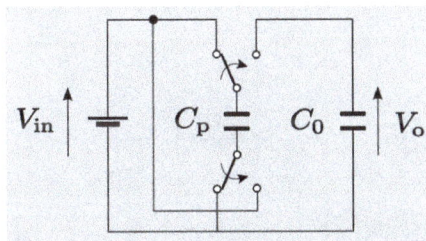

Charge-pump voltage doubler schematic

Another basic concept is the charge pump, a version of which is shown schematically in the above figure. The charge pump capacitor, C_p, is first charged to the input voltage. It is then switched to charging the output capacitor, C_o, in series with the input voltage resulting in C_o eventually being charged to twice the input voltage. It may take several cycles before the charge pump succeeds in fully charging C_o but after steady state has been reached it is only necessary for C_p to pump a small amount of charge equivalent to that being supplied to the load from C_o. While C_o is disconnected from the charge pump it partially discharges into the load resulting in ripple on the output voltage. This ripple is smaller for higher clock frequencies since the discharge time is shorter, and is also easier to filter. Alternatively, the capacitors can be made smaller for a given ripple specification. The practical maximum clock frequency in integrated circuits is typically in the hundreds of kilohertz.

Dickson Charge Pump

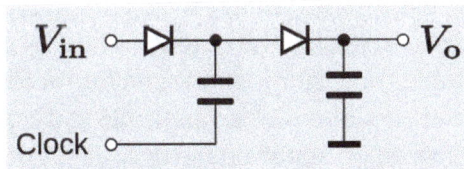

Dickson charge-pump voltage-doubler

The Dickson charge pump, or Dickson multiplier, consists of a cascade of diode/capacitor cells with the bottom plate of each capacitor driven by a clock pulse train. The circuit is a modification of the Cockcroft-Walton multiplier but takes a DC input with the clock trains providing the switching signal instead of the AC input. The Dickson multiplier normally requires that alternate cells are driven from clock pulses of opposite phase. However, since a voltage doubler, shown in the above figure, requires only one stage of multiplication only one clock signal is required.

The Dickson multiplier is frequently employed in integrated circuits where the supply voltage (from a battery for instance) is lower than that required by the circuitry. It is advantageous in integrated circuit manufacture that all the semiconductor components are of basically the same type. MOSFETs are commonly the standard logic block in many integrated circuits. For this reason the diodes are often replaced by this type of transistor, but wired to function as a diode - an arrangement called a diode-wired MOSFET. Figure below shows a Dickson voltage doubler using diode-wired n-channel enhancement type MOSFETs.

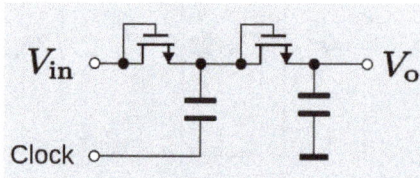

Dickson voltage doubler using diode-wired MOSFETs

There are many variations and improvements to the basic Dickson charge pump. Many of these are concerned with reducing the effect of the transistor drain-source voltage. This can be very significant if the input voltage is small, such as a low-voltage battery. With ideal switching elements the output is an integral multiple of the input (two for a doubler) but with a single-cell battery as the input source and MOSFET switches the output will be far less than this value since much of the voltage will be dropped across the transistors. For a circuit using discrete components the Schottky diode would be a better choice of switching element for its extremely low voltage drop in the on state. However, integrated circuit designers prefer to use the easily available MOSFET and compensate for its inadequacies with increased circuit complexity.

As an example, an alkaline battery cell has a nominal voltage of 1.5 V. A voltage doubler using ideal switching elements with zero voltage drop will output double this, namely 3.0 V. However, the drain-source voltage drop of a diode-wired MOSFET when it is in the on state must be at least the gate threshold voltage which might typically be 0.9 V. This voltage "doubler" will only succeed in raising the output voltage by about 0.6 V to 2.1 V. If the drop across the final smoothing transistor is also taken into account the circuit may not be able to increase the voltage at all without using multiple stages. A typical Schottky diode, on the other hand, might have an on state voltage of 0.3 V. A doubler using this Schottky diode will result in a voltage of 2.7 V, or at the output after the smoothing diode, 2.4 V.

Cross-coupled Switched Capacitors

Cross-coupled switched capacitor circuits come into their own for very low input voltages. Wireless battery driven equipment such as pagers, bluetooth devices and the like may require a single-cell battery to continue to supply power when it has discharged to under a volt.

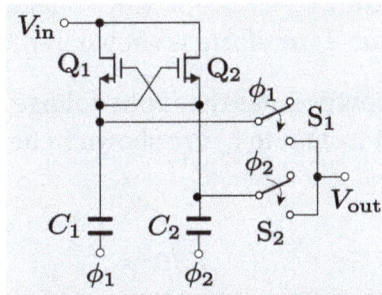

Cross-coupled switched-capacitor voltage doubler

When clock ϕ_1 is low transistor Q_2 is turned off. At the same time clock ϕ_2 is high turning on transistor Q_1 resulting in capacitor C_1 being charged to V_{in}. When ϕ_1 goes high the top plate of C_1 is pushed up to twice V_{in}. At the same time switch S_1 closes so this voltage appears at the output. At the same time Q_2 is turned on allowing C_2 to charge. On the next half cycle the roles will be reversed: ϕ_1 will be low, ϕ_2 will be high, S_1 will open and S_2 will close. Thus, the output is supplied with $2V_{in}$ alternately from each side of the circuit.

The loss is low in this circuit because there are no diode-wired MOSFETs and their associated threshold voltage problems. The circuit also has the advantage that the ripple frequency is doubled because there are effectively two voltage doublers both supplying the output from out of phase clocks. The primary disadvantage of this circuit is that stray capacitances are much more significant than with the Dickson multiplier and account for the larger part of the losses in this circuit.

Voltage Doubler

A voltage doubler circuit is shown in the left figure. The circuit produces a dc voltage, which is double the peak input voltage.

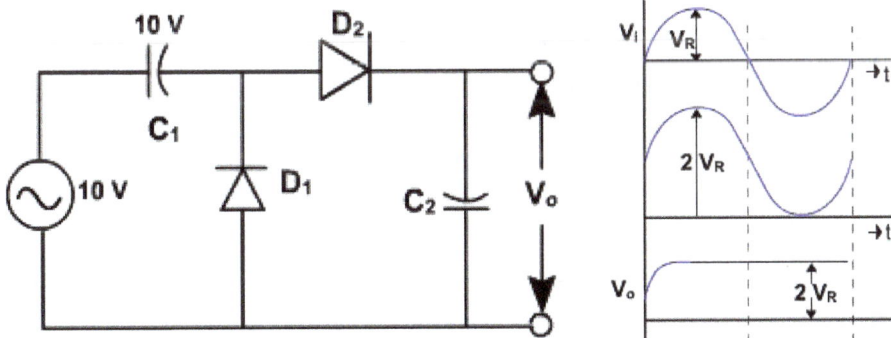

At the peak of the negative half cycle D_1 is forward based, and D_2 is reverse based. This charges C_1 to the peak voltage V_p with the polarity shown. At the peak of the positive half cycle D_1 is reverse biased and D_2 is forward biased. Because the source and C_1 are in series, C_2 will change toward $2V_p$. e.g. Capacitor voltage increases continuously and finally becomes 20V. The voltage waveform is shown in the right figure.

To understand the circuit operation, let the input voltage varies from -10 V to +10 V. The different stages of circuit from 0 to t_{10} are shown in fig. (a).

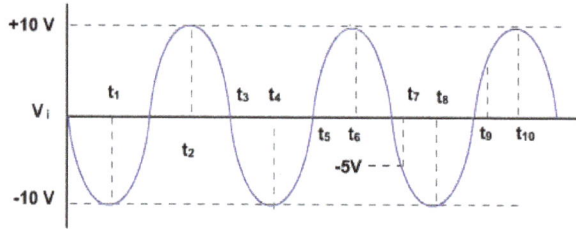

Fig. (a)

During 0 to t_1, the input voltage is negative, D_1 is forward biased the capacitor is charged to −10 V with the polarity as shown in fig. b.

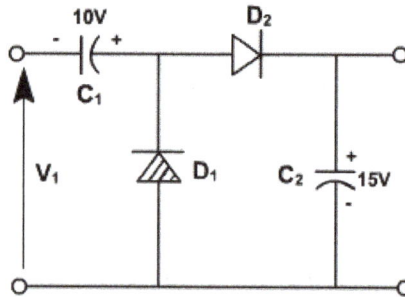

Fig. (b)

During t_1 to t_2, D_2 becomes forward biased and conducts and at t_2, when V_i is 10V total voltage change is 20V. If $C_1 = C_2 = C$, both the capacitor voltages charge to +10 V i.e. C_1 voltage becomes 0 and C_2 charges to +10V.

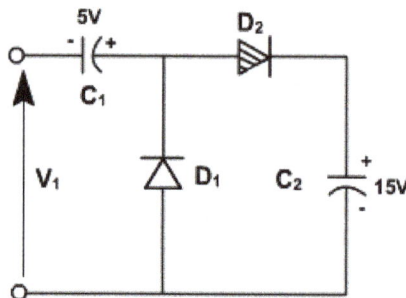

Fig. (c)

From t_2 to t_3 there is no conduction as both D_1 and D_2 are reverse biased.

During t_3 to t_4 D_1 is forward biased and conducts. C_1 again charges to +10V

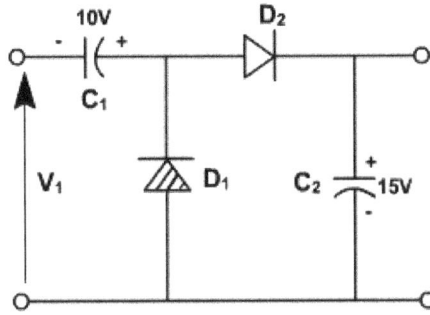

Fig. (d)

During t_4 to t_5 both D_1 and D_2 are reverse biased and do not conduct.

During t_5 to t_6 D_2 is forward biased and conducts. The capacitor C_2 voltage becomes +15 V and C_1 voltage becomes +5 V.

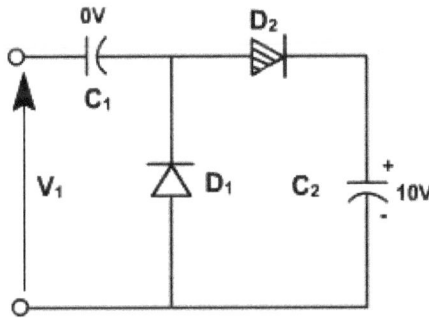

Fig. (e)

Again during t_6 to t_7 there is no conduction and during t_7 to t_8, D_1 conducts. The capacitor C_1 recharges to 10 V.

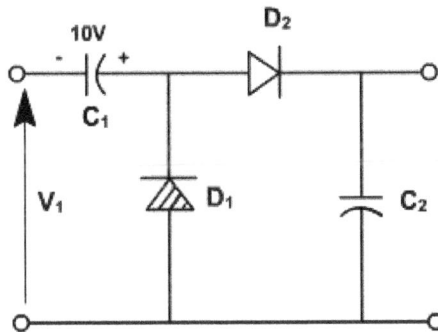

Fig. (f)

During t_8 to t_9 both D_1 and D_2 are reverse biased and there is no conduction. During t_9 to t_{10} D_2 conducts and capacitor C_2 voltage becomes + 17.5 V and C_1 voltage becomes 7.5V. This process continues till the capacitor C_1 voltage becomes +20V.

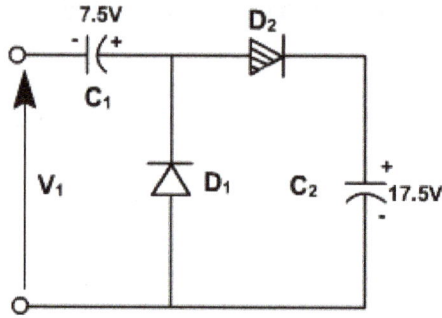

Fig. (g)

References

- Clarence W. De Silva, Vibration monitoring, testing, and instrumentation, pp.2.43-2.49, CRC Press, 2007 ISBN 1-4200-5319-1

- Comer, Donald T. (1996). "Zener Zap Anti-Fuse Trim in VLSI Circuits". VLSI Design. 5: 89. doi:10.1155/1996/23706

- Martin Hartley Jones (1995). A Practical Introduction to Electronic Circuits. Cambridge University Press. p. 261. ISBN 978-0-521-47879-3

- Horowitz, Paul; Hill, Winfield (1989). The Art of Electronics (2nd ed.). Cambridge University Press. pp. 68–69. ISBN 0-521-37095-7

- Diffenderfer, Robert (2005). Electronic Devices: Systems and Applications. Thomas Delmar Learning. pp. 95–100. ISBN 1401835147. Retrieved July 22, 2014

- Dorf, Richard C., ed. (1993). The Electrical Engineering Handbook. Boca Raton: CRC Press. p. 457. ISBN 0-8493-0185-8

3

Transistors and its Types

A transistor is a semiconductor that is used to increase and switch electronic power and signal. It is usually made up of silicon or germanium and has three separate terminals. In this section, two major types of transistors are discussed. They are bipolar junction transistors and unijunction transistors. The chapter serves as a source to understand the major categories related to transistors.

Transistor

Assorted discrete transistors. Packages in order from top to bottom: TO-3, TO-126, TO-92, SOT-23.

A transistor is a semiconductor device used to amplify or switch electronic signals and electrical power. It is composed of semiconductor material usually with at least three terminals for connection to an external circuit. A voltage or current applied to one pair of the transistor's terminals controls the current through another pair of terminals. Because the controlled (output) power can be higher than the controlling (input) power, a transistor can amplify a signal. Today, some transistors are packaged individually, but many more are found embedded in integrated circuits.

The transistor is the fundamental building block of modern electronic devices, and is ubiquitous in modern electronic systems. Julius Edgar Lilienfeld patented a field-effect transistor in 1926 but it was not possible to actually construct a working device at that time. The first practically implemented device was a point-contact transistor invented in 1947 by American physicists John Bardeen, Walter Brattain, and William Shockley. The transistor revolutionized the field of electronics, and paved the way for smaller and cheaper radios, calculators, and computers, among other things. The transistor is on the list of IEEE milestones in electronics, and Bardeen, Brattain, and Shockley shared the 1956 Nobel Prize in Physics for their achievement.

History

A replica of the first working transistor.

The thermionic triode, a vacuum tube invented in 1907, enabled amplified radio technology and long-distance telephony. The triode, however, was a fragile device that consumed a substantial amount of power. In 1909 physicist William Eccles discovered the crystal diode oscillator. Physicist Julius Edgar Lilienfeld filed a patent for a field-effect transistor (FET) in Canada in 1925, which was intended to be a solid-state replacement for the triode. Lilienfeld also filed identical patents in the United States in 1926 and 1928. However, Lilienfeld did not publish any research articles about his devices nor did his patents cite any specific examples of a working prototype. Because the production of high-quality semiconductor materials was still decades away, Lilienfeld's solid-state amplifier ideas would not have found practical use in the 1920s and 1930s, even if such a device had been built. In 1934, German inventor Oskar Heil patented a similar device in Europe.

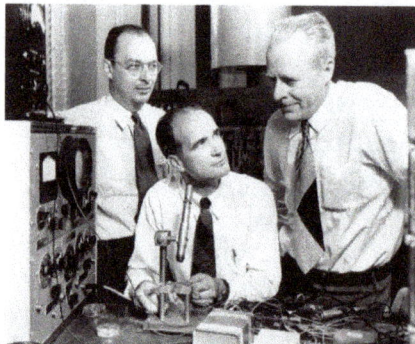

John Bardeen, William Shockley and Walter Brattain at Bell Labs.

From November 17, 1947 to December 23, 1947, John Bardeen and Walter Brattain at AT&T's Bell Labs in Murray Hill, New Jersey of the United States performed experiments and observed that when two gold point contacts were applied to a crystal of germanium, a signal was produced with the output power greater than the input. Solid State Physics Group leader William Shockley saw the potential in this, and over the next few months worked to greatly expand the knowledge of semiconductors. The term *transistor* was coined by John R. Pierce as a contraction of the term *transresistance*. According to Lillian Hoddeson and Vicki Daitch, authors of a biography of John Bardeen, Shockley had proposed that Bell Labs' first patent for a transistor should be based on the field-effect and that he be named as the inventor. Having unearthed Lilienfeld's patents that went into obscurity years earlier, lawyers at Bell Labs advised against Shockley's proposal because the idea of a field-effect transistor that used an electric field as a "grid" was not new. Instead, what Bardeen, Brattain, and Shockley invented in 1947 was the first point-contact transistor. In acknowledgement of this accomplishment, Shockley, Bardeen, and Brattain were jointly awarded the 1956 Nobel Prize in Physics "for their researches on semiconductors and their discovery of the transistor effect".

Herbert F. Mataré (1950)

In 1948, the point-contact transistor was independently invented by German physicists Herbert Mataré and Heinrich Welker while working at the *Compagnie des Freins et Signaux*, a Westinghouse subsidiary located in Paris. Mataré had previous experience in developing crystal rectifiers from silicon and germanium in the German radar effort during World War II. Using this knowledge, he began researching the phenomenon of "interference" in 1947. By June 1948, witnessing currents flowing through point-contacts, Mataré produced consistent results using samples of germanium produced by Welker, similar to what Bardeen and Brattain had accomplished earlier in December 1947. Realizing that Bell Labs' scientists had already invented the transistor before them, the company rushed to get its "transistron" into production for amplified use in France's telephone network.

The first bipolar junction transistors were invented by Bell labs William Shockley, which applied for patent(2,569,347) on June 26, 1948. On April 12, 1950, Bell Labs chemists Gordon Teal and Morgan Sparks had successfully produced a working bipolar NPN junction amplifying germanium transistor. Bell Labs had made this new "sandwich" transistor discovery announcement, in a press release on July 4, 1951.

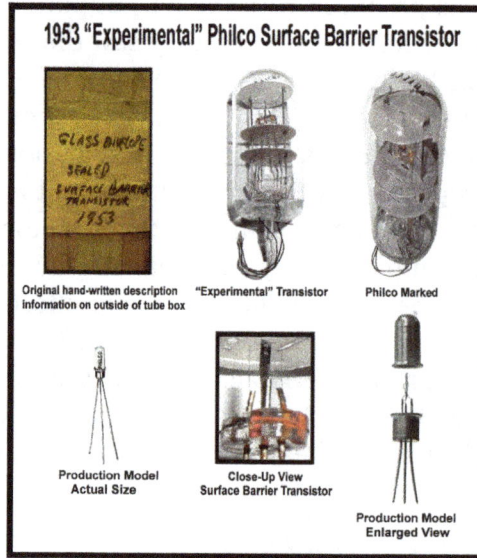

Philco surface-barrier transistor developed and produced in 1953

The first high-frequency transistor was the surface-barrier germanium transistor developed by Philco in 1953, capable of operating up to 60 MHz. These were made by etching depressions into an N-type germanium base from both sides with jets of Indium(III) sulfate until it was a few ten-thousandths of an inch thick. Indium electroplated into the depressions formed the collector and emitter.

The first "prototype" pocket transistor radio was shown by INTERMETALL (a company founded by Herbert Mataré in 1952) at the *Internationale Funkausstellung Düsseldorf* between August 29, 1953 and September 9, 1953.

The first "production" all-transistor radio was the Regency TR-1, released in October 1954. Produced as a joint venture between the Regency Division of Industrial Development Engineering Associates, I.D.E.A. and Texas Instruments of Dallas Texas, the TR-1 was manufactured in Indianapolis, Indiana. It was a near pocket-sized radio featuring 4 transistors and one germanium diode. The industrial design was outsourced to the Chicago firm of industrial design firm of Painter, Teague and Petertil. It was initially released in one of four different colours: black, bone white, red, and gray. Other colours were to shortly follow.

The first working silicon transistor was developed at Bell Labs on January 26, 1954 by Morris Tanenbaum. The first commercial silicon transistor was produced by Texas Instruments in 1954. This was the work of Gordon Teal, an expert in growing crystals of high purity, who had previously worked at Bell Labs. The first MOSFET actually built was by Kahng and Atalla at Bell Labs in 1960.

Importance

A Darlington transistor opened up so the actual transistor chip (the small square) can be seen inside. A Darlington transistor is effectively two transistors on the same chip. One transistor is much larger than the other, but both are large in comparison to transistors in large-scale integration because this particular example is intended for power applications.

The transistor is the key active component in practically all modern electronics. Many consider it to be one of the greatest inventions of the 20th century. Its importance in today's society rests on its ability to be mass-produced using a highly automated process (semiconductor device fabrication) that achieves astonishingly low per-transistor costs. The invention of the first transistor at Bell Labs was named an IEEE Milestone in 2009.

Although several companies each produce over a billion individually packaged (known as *discrete*) transistors every year, the vast majority of transistors are now produced in integrated circuits (often shortened to *IC*, *microchips* or simply *chips*), along with diodes, resistors, capacitors and other electronic components, to produce complete electronic circuits. A logic gate consists of up to about twenty transistors whereas an advanced microprocessor, as of 2009, can use as many as 3 billion transistors (MOS-FETs). "About 60 million transistors were built in 2002... for [each] man, woman, and child on Earth."

The transistor's low cost, flexibility, and reliability have made it a ubiquitous device. Transistorized mechatronic circuits have replaced electromechanical devices in controlling appliances and machinery. It is often easier and cheaper to use a standard microcontroller and write a computer program to carry out a control function than to design an equivalent mechanical system to control that same function.

Simplified Operation

The essential usefulness of a transistor comes from its ability to use a small signal applied between one pair of its terminals to control a much larger signal at another pair of terminals. This property is called gain. It can produce a stronger output signal, a voltage or current, which is proportional to a weaker input signal; that is, it can act as an amplifier. Alternatively, the transistor can be used to turn current on or off in a circuit as an electrically controlled switch, where the amount of current is determined by other circuit elements.

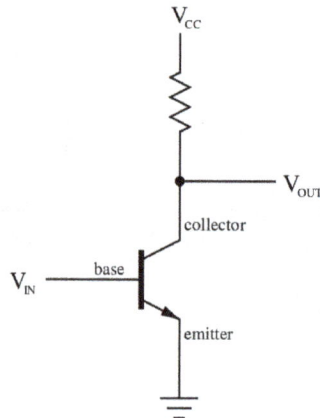

A simple circuit diagram to show the labels of a n–p–n bipolar transistor.

There are two types of transistors, which have slight differences in how they are used in a circuit. A *bipolar transistor* has terminals labeled base, collector, and emitter. A small current at the base terminal (that is, flowing between the base and the emitter) can control or switch a much larger current between the collector and emitter terminals. For a *field-effect transistor*, the terminals are labeled gate, source, and drain, and a voltage at the gate can control a current between source and drain.

The image represents a typical bipolar transistor in a circuit. Charge will flow between emitter and collector terminals depending on the current in the base. Because internally the base and emitter connections behave like a semiconductor diode, a voltage drop develops between base and emitter while the base current exists. The amount of this voltage depends on the material the transistor is made from, and is referred to as V_{BE}.

Transistor as a Switch

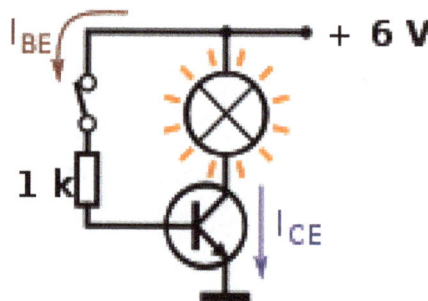

BJT used as an electronic switch, in grounded-emitter configuration.

Transistors are commonly used in digital circuits as electronic switches which can be either in an "on" or "off" state, both for high-power applications such as switched-mode power supplies and for low-power applications such as logic gates. Important parameters for this application include the current switched, the voltage handled, and the switching speed, characterised by the rise and fall times.

In a grounded-emitter transistor circuit, such as the light-switch circuit shown, as the base voltage rises, the emitter and collector currents rise exponentially. The collector voltage drops because of reduced resistance from collector to emitter. If the voltage difference between the collector and emitter were zero (or near zero), the collector current would be limited only by the load resistance (light bulb) and the supply voltage. This is called *saturation* because current is flowing from collector to emitter freely. When saturated, the switch is said to be *on*.

Providing sufficient base drive current is a key problem in the use of bipolar transistors as switches. The transistor provides current gain, allowing a relatively large current in the collector to be switched by a much smaller current into the base terminal. The ratio of these currents varies depending on the type of transistor, and even for a particular type, varies depending on the collector current. In the example light-switch circuit shown, the resistor is chosen to provide enough base current to ensure the transistor will be saturated.

In a switching circuit, the idea is to simulate, as near as possible, the ideal switch having the properties of open circuit when off, short circuit when on, and an instantaneous transition between the two states. Parameters are chosen such that the "off" output is limited to leakage currents too small to affect connected circuitry; the resistance of the transistor in the "on" state is too small to affect circuitry; and the transition between the two states is fast enough not to have a detrimental effect.

Transistor as an Amplifier

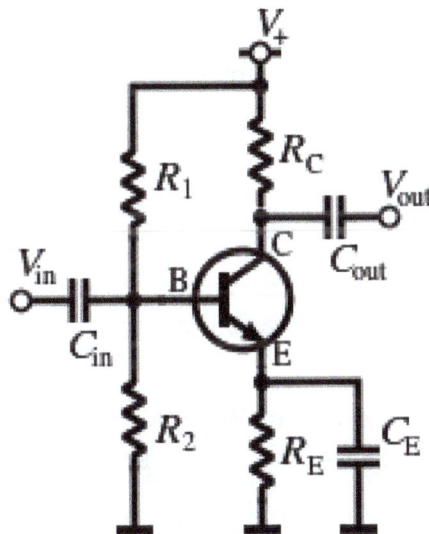

Amplifier circuit, common-emitter configuration with a voltage-divider bias circuit.

The common-emitter amplifier is designed so that a small change in voltage (V_{in}) changes the small current through the base of the transistor; the transistor's current amplification combined with the properties of the circuit means that small swings in V_{in} produce large changes in V_{out}.

Various configurations of single transistor amplifier are possible, with some providing current gain, some voltage gain, and some both.

From mobile phones to televisions, vast numbers of products include amplifiers for sound reproduction, radio transmission, and signal processing. The first discrete-transistor audio amplifiers barely supplied a few hundred milliwatts, but power and audio fidelity gradually increased as better transistors became available and amplifier architecture evolved.

Modern transistor audio amplifiers of up to a few hundred watts are common and relatively inexpensive.

Comparison with Vacuum Tubes

Before transistors were developed, vacuum (electron) tubes (or in the UK "thermionic valves" or just "valves") were the main active components in electronic equipment.

Advantages

The key advantages that have allowed transistors to replace vacuum tubes in most applications are:

- no cathode heater (which produces the characteristic orange glow of tubes), reducing power consumption, eliminating delay as tube heaters warm up, and immune from cathode poisoning and depletion;

- very small size and weight, reducing equipment size;

- large numbers of extremely small transistors can be manufactured as a single integrated circuit;

- low operating voltages compatible with batteries of only a few cells;

- circuits with greater energy efficiency are usually possible. For low-power applications (e.g., voltage amplification) in particular, energy consumption can be very much less than for tubes;

- inherent reliability and very long life; tubes always degrade and fail over time. Some transistorized devices have been in service for more than 50 years;

- complementary devices available, providing design flexibility including complementary-symmetry circuits, not possible with vacuum tubes;

- very low sensitivity to mechanical shock and vibration, providing physical ruggedness and virtually eliminating shock-induced spurious signals (e.g., microphonics in audio applications);

- not susceptible to breakage of a glass envelope, leakage, outgassing, and other physical damage.

Limitations

Transistors have the following limitations:

- silicon transistors can age and fail;

- high-power, high-frequency operation, such as that used in over-the-air television broadcasting, is better achieved in vacuum tubes due to improved electron mobility in a vacuum;

- solid-state devices are susceptible to damage from very brief electrical and thermal events, including electrostatic discharge in handling; vacuum tubes are electrically much more rugged;

- sensitivity to radiation and cosmic rays (special radiation-hardened chips are used for spacecraft devices);

- vacuum tubes in audio applications create significant lower-harmonic distortion, the so-called tube sound, which some people prefer.

Types

BJT and JFET symbols

JFET	MOSFET enh		MOSFET dep	N-channel

JFET and MOSFET symbols

Transistors are categorized by:

- semiconductor material: the metalloids germanium (first used in 1947) and silicon (first used in 1954)—in amorphous, polycrystalline and monocrystalline form—, the compounds gallium arsenide (1966) and silicon carbide (1997), the alloy silicon-germanium (1989), the allotrope of carbon graphene (research ongoing since 2004), etc.

- structure: BJT, JFET, IGFET (MOSFET), insulated-gate bipolar transistor, "other types";

- electrical polarity (positive and negative): n–p–n, p–n–p (BJTs), n-channel, p-channel (FETs);

- maximum power rating: low, medium, high;

- maximum operating frequency: low, medium, high, radio (RF), microwave frequency (the maximum effective frequency of a transistor in a common-emitter or common-source circuit is denoted by the term f_T, an abbreviation for transition frequency—the frequency of transition is the frequency at which the transistor yields unity voltage gain)

- application: switch, general purpose, audio, high voltage, super-beta, matched pair;

- physical packaging: through-hole metal, through-hole plastic, surface mount, ball grid array, power modules;

- amplification factor h_{FE}, β_F (transistor beta) or g_m (transconductance).

Hence, a particular transistor may be described as *silicon, surface-mount, BJT, n–p–n, low-power, high-frequency switch.*

A popular way to remember which symbol represents which type of transistor is to look at the arrow and how it is arranged. Within an NPN transistor symbol, the arrow will Not Point iN. Conversely, within the PNP symbol you see that the arrow Points iN Proudly.

Bipolar Junction Transistor (BJT)

Bipolar transistors are so named because they conduct by using both majority and minority carriers. The bipolar junction transistor, the first type of transistor to be mass-produced, is a combination of two junction diodes, and is formed of either a thin layer of p-type semiconductor sandwiched between two n-type semiconductors (an n–p–n transistor), or a thin layer of n-type semiconductor sandwiched between two p-type semiconductors (a p–n–p transistor). This construction produces two p–n junctions: a base–emitter junction and a base–collector junction, separated by a thin region of semiconductor known as the base region (two junction diodes wired together without sharing an intervening semiconducting region will not make a transistor).

BJTs have three terminals, corresponding to the three layers of semiconductor—an *emitter*, a *base*, and a *collector*. They are useful in amplifiers because the currents at the emitter and collector are controllable by a relatively small base current. In an n–p–n transistor operating in the active region, the emitter–base junction is forward biased (electrons and holes recombine at the junction), and electrons are injected into the base region. Because the base is narrow, most of these electrons will diffuse into the reverse-biased (electrons and holes are formed at, and move away from the junction) base–collector junction and be swept into the collector; perhaps one-hundredth of the electrons will recombine in the base, which is the dominant mechanism in the base current. By controlling the number of electrons that can leave the base, the number of electrons entering the collector can be controlled. Collector current is approximately β (common-emitter current gain) times the base current. It is typically greater than 100 for small-signal transistors but can be smaller in transistors designed for high-power applications.

Unlike the field-effect transistor, the BJT is a low-input-impedance device. Also, as the base–emitter voltage (V_{BE}) is increased the base–emitter current and hence the collector–emitter current (I_{CE}) increase exponentially according to the Shockley diode model and the Ebers-Moll model. Because of this exponential relationship, the BJT has a higher transconductance than the FET.

Bipolar transistors can be made to conduct by exposure to light, because absorption of photons in the base region generates a photocurrent that acts as a base current; the collector current is approximately β times the photocurrent. Devices designed for this purpose have a transparent window in the package and are called phototransistors.

Field-effect Transistor (FET)

The *field-effect transistor*, sometimes called a *unipolar transistor*, uses either electrons (in *n-channel FET*) or holes (in *p-channel FET*) for conduction. The four terminals of the FET are named *source*, *gate*, *drain*, and *body* (*substrate*). On most FETs, the body is connected to the source inside the package, and this will be assumed for the following description.

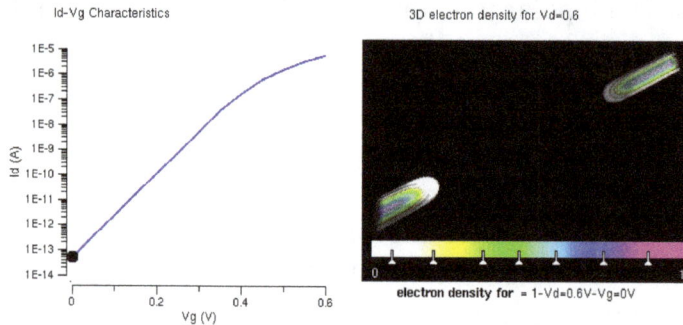

Operation of a FET and its Id-Vg curve. At first, when no gate voltage is applied. There is no inversion electron in the channel, the device is OFF. As gate voltage increase, inversion electron density in the channel increase, current increase, the device turns on.

In a FET, the drain-to-source current flows via a conducting channel that connects the *source* region to the *drain* region. The conductivity is varied by the electric field that is produced when a voltage is applied between the gate and source terminals; hence the current flowing between the drain and source is controlled by the voltage applied between the gate and source. As the gate–source voltage (V_{GS}) is increased, the drain–source current (I_{DS}) increases exponentially for V_{GS} below threshold, and then at a roughly quadratic rate ($I_{GS} \propto (V_{GS} - V_T)^2$) (where V_T is the threshold voltage at which drain current begins) in the "space-charge-limited" region above threshold. A quadratic behavior is not observed in modern devices, for example, at the 65 nm technology node.

For low noise at narrow bandwidth the higher input resistance of the FET is advantageous.

FETs are divided into two families: *junction FET* (JFET) and *insulated gate FET* (IGFET). The IGFET is more commonly known as a *metal–oxide–semiconductor FET* (MOSFET), reflecting its original construction from layers of metal (the gate), oxide (the insulation), and semiconductor. Unlike IGFETs, the JFET gate forms a p–n diode with the channel which lies between the source and drain. Functionally, this makes the n-channel JFET the solid-state equivalent of the vacuum tube triode which, similarly, forms a diode between its grid and cathode. Also, both devices operate in the *depletion mode*, they both have a high input impedance, and they both conduct current under the control of an input voltage.

Metal–semiconductor FETs (MESFETs) are JFETs in which the reverse biased p–n junction is replaced by a metal–semiconductor junction. These, and the HEMTs (high-electron-mobility transistors, or HFETs), in which a two-dimensional electron gas with very high carrier mobility is used for charge transport, are especially suitable for use at very high frequencies (microwave frequencies; several GHz).

FETs are further divided into *depletion-mode* and *enhancement-mode* types, depending on whether the channel is turned on or off with zero gate-to-source voltage. For enhancement mode, the channel is off at zero bias, and a gate potential can "enhance" the conduction. For the depletion mode, the channel is on at zero bias, and a gate potential (of the opposite polarity) can "deplete" the channel, reducing conduction. For either mode, a more positive gate voltage corresponds to a higher current for n-channel devices and a lower current for p-channel devices. Nearly all JFETs are depletion-mode because the diode junctions would forward bias and conduct if they were enhancement-mode devices; most IGFETs are enhancement-mode types.

Usage of Bipolar and Field-effect Transistors

The bipolar junction transistor (BJT) was the most commonly used transistor in the 1960s and 70s. Even after MOSFETs became widely available, the BJT remained the transistor of choice for many analog circuits such as amplifiers because of their greater linearity and ease of manufacture. In integrated circuits, the desirable properties of MOSFETs allowed them to capture nearly all market share for digital circuits. Discrete MOSFETs can be applied in transistor applications, including analog circuits, voltage regulators, amplifiers, power transmitters and motor drivers.

Other Transistor Types

Transistor symbol created on Portuguese pavement in the University of Aveiro.

- Bipolar junction transistor (BJT):

  o heterojunction bipolar transistor, up to several hundred GHz, common in modern ultrafast and RF circuits;

  o Schottky transistor;

  o avalanche transistor;

  o Darlington transistors are two BJTs connected together to provide a high current gain equal to the product of the current gains of the two transistors;

o   insulated-gate bipolar transistors (IGBTs) use a medium-power IGFET, similarly connected to a power BJT, to give a high input impedance. Power diodes are often connected between certain terminals depending on specific use. IGBTs are particularly suitable for heavy-duty industrial applications. The ASEA Brown Boveri (ABB) *5SNA2400E170100* illustrates just how far power semiconductor technology has advanced. Intended for three-phase power supplies, this device houses three n–p–n IGBTs in a case measuring 38 by 140 by 190 mm and weighing 1.5 kg. Each IGBT is rated at 1,700 volts and can handle 2,400 amperes;

o   phototransistor;

o   multiple-emitter transistor, used in transistor–transistor logic and integrated current mirrors;

o   multiple-base transistor, used to amplify very-low-level signals in noisy environments such as the pickup of a record player or radio front ends. Effectively, it is a very large number of transistors in parallel where, at the output, the signal is added constructively, but random noise is added only stochastically.

• Field-effect transistor (FET):

o   carbon nanotube field-effect transistor (CNFET), where the channel material is replaced by a carbon nanotube;

o   junction gate field-effect transistor (JFET), where the gate is insulated by a reverse-biased p–n junction;

o   metal–semiconductor field-effect transistor (MESFET), similar to JFET with a Schottky junction instead of a p–n junction;

• high-electron-mobility transistor (HEMT);

o   metal–oxide–semiconductor field-effect transistor (MOSFET), where the gate is insulated by a shallow layer of insulator;

o   inverted-T field-effect transistor (ITFET);

o   fin field-effect transistor (FinFET), source/drain region shapes fins on the silicon surface;

o   fast-reverse epitaxial diode field-effect transistor (FREDFET);

o   thin-film transistor, in LCDs;

o   organic field-effect transistor (OFET), in which the semiconductor is an organic compound;

- o ballistic transistor;

- o floating-gate transistor, for non-volatile storage;

- o FETs used to sense environment;

- ion-sensitive field-effect transistor (IFSET), to measure ion concentrations in solution,

- electrolyte–oxide–semiconductor field-effect transistor (EOSFET), neurochip,

- deoxyribonucleic acid field-effect transistor (DNAFET).

- Tunnel field-effect transistor, where it switches by modulating quantum tunnelling through a barrier.

- Diffusion transistor, formed by diffusing dopants into semiconductor substrate; can be both BJT and FET.

- Unijunction transistor, can be used as simple pulse generators. It comprise a main body of either P-type or N-type semiconductor with ohmic contacts at each end (terminals *Base1* and *Base2*). A junction with the opposite semiconductor type is formed at a point along the length of the body for the third terminal (*Emitter*).

- Single-electron transistors (SET), consist of a gate island between two tunneling junctions. The tunneling current is controlled by a voltage applied to the gate through a capacitor.

- Nanofluidic transistor, controls the movement of ions through sub-microscopic, water-filled channels.

- Multigate devices:

  - o tetrode transistor;

  - o pentode transistor;

  - o trigate transistor (prototype by Intel);

  - o dual-gate field-effect transistors have a single channel with two gates in cascode; a configuration optimized for *high-frequency amplifiers*, *mixers*, and oscillators.

- Junctionless nanowire transistor (JNT), uses a simple nanowire of silicon surrounded by an electrically isolated "wedding ring" that acts to gate the flow of electrons through the wire.

- Vacuum-channel transistor, when in 2012, NASA and the National Nanofab Center in South Korea were reported to have built a prototype vacuum-channel transistor in only 150 nanometers in size, can be manufactured cheaply using standard silicon semiconductor processing, can operate at high speeds even in hostile environments, and could consume just as much power as a standard transistor.

- Organic electrochemical transistor.

Part Numbering Standards/Specifications

The types of some transistors can be parsed from the part number. There are three major semiconductor naming standards; in each the alphanumeric prefix provides clues to type of the device.

Japanese Industrial Standard (JIS)

JIS Transistor Prefix Table	
Prefix	Type of transistor
2SA	high-frequency p–n–p BJTs
2SB	audio-frequency p–n–p BJTs
2SC	high-frequency n–p–n BJTs
2SD	audio-frequency n–p–n BJTs
2SJ	P-channel FETs (both JFETs and MOSFETs)
2SK	N-channel FETs (both JFETs and MOSFETs)

The *JIS-C-7012* specification for transistor part numbers starts with "2S", e.g. 2SD965, but sometimes the "2S" prefix is not marked on the package – a 2SD965 might only be marked "D965"; a 2SC1815 might be listed by a supplier as simply "C1815". This series sometimes has suffixes (such as "R", "O", "BL", standing for "red", "orange", "blue", etc.) to denote variants, such as tighter h_{FE} (gain) groupings.

European Electronic Component Manufacturers Association (EECA)

The Pro Electron standard, the European Electronic Component Manufacturers Association part numbering scheme, begins with two letters: the first gives the semiconductor type (A for germanium, B for silicon, and C for materials like GaAs); the second letter denotes the intended use (A for diode, C for general-purpose transistor, etc.). A 3-digit sequence number (or one letter then 2 digits, for industrial types) follows. With early devices this indicated the case type. Suffixes may be used, with a letter (e.g. "C" often means high h_{FE}, such as in: BC549C) or other codes may follow to show gain (e.g. BC327-25) or voltage rating (e.g. BUK854-800A). The more common prefixes are:

Pro Electron / EECA Transistor Prefix Table			
Prefix class	Type and usage	Example	Equivalent
AC	Germanium small-signal AF transistor	AC126	NTE102A
AD	Germanium AF power transistor	AD133	NTE179
AF	Germanium small-signal RF transistor	AF117	NTE160
AL	Germanium RF power transistor	ALZ10	NTE100
AS	Germanium switching transistor	ASY28	NTE101
AU	Germanium power switching transistor	AU103	NTE127
BC	Silicon, small-signal transistor ("general purpose")	BC548	2N3904
BD	Silicon, power transistor	BD139	NTE375
BF	Silicon, RF (high frequency) BJT or FET	BF245	NTE133
BS	Silicon, switching transistor (BJT or MOSFET)	BS170	2N7000
BL	Silicon, high frequency, high power (for transmitters)	BLW60	NTE325
BU	Silicon, high voltage (for CRT horizontal deflection circuits)	BU2520A	NTE2354
CF	Gallium Arsenide small-signal Microwave transistor (MESFET)	CF739	—
CL	Gallium Arsenide Microwave power transistor (FET)	CLY10	—

Joint Electron Device Engineering Council (JEDEC)

The JEDEC *EIA370* transistor device numbers usually start with "2N", indicating a three-terminal device (dual-gate field-effect transistors are four-terminal devices, so begin with 3N), then a 2, 3 or 4-digit sequential number with no significance as to device properties (although early devices with low numbers tend to be germanium). For example, 2N3055 is a silicon n–p–n power transistor, 2N1301 is a p–n–p germanium switching transistor. A letter suffix (such as "A") is sometimes used to indicate a newer variant, but rarely gain groupings.

Proprietary

Manufacturers of devices may have their own proprietary numbering system, for example CK722. Since devices are second-sourced, a manufacturer's prefix (like "MPF" in MPF102, which originally would denote a Motorola FET) now is an unreliable indicator of who made the device. Some proprietary naming schemes adopt parts of other naming schemes, for example a PN2222A is a (possibly Fairchild Semiconductor) 2N2222A in a plastic case (but a PN108 is a plastic version of a BC108, not a 2N108, while the PN100 is unrelated to other xx100 devices).

Military part numbers sometimes are assigned their own codes, such as the British Military CV Naming System.

Manufacturers buying large numbers of similar parts may have them supplied with "house numbers", identifying a particular purchasing specification and not necessarily a device with a standardized registered number. For example, an HP part 1854,0053 is a (JEDEC) 2N2218 transistor which is also assigned the CV number: CV7763.

Naming Problems

With so many independent naming schemes, and the abbreviation of part numbers when printed on the devices, ambiguity sometimes occurs. For example, two different devices may be marked "J176" (one the J176 low-power JFET, the other the higher-powered MOSFET 2SJ176).

As older "through-hole" transistors are given surface-mount packaged counterparts, they tend to be assigned many different part numbers because manufacturers have their own systems to cope with the variety in pinout arrangements and options for dual or matched n–p–n+p–n–p devices in one pack. So even when the original device (such as a 2N3904) may have been assigned by a standards authority, and well known by engineers over the years, the new versions are far from standardized in their naming.

Construction

Semiconductor Material

Semiconductor material characteristics				
Semiconductor material	Junction forward voltage V @ 25 °C	Electron mobility m²/(V·s) @ 25 °C	Hole mobility m²/(V·s) @ 25 °C	Max. junction temp. °C
Ge	0.27	0.39	0.19	70 to 100
Si	0.71	0.14	0.05	150 to 200
GaAs	1.03	0.85	0.05	150 to 200
Al–Si junction	0.3	—	—	150 to 200

The first BJTs were made from germanium (Ge). Silicon (Si) types currently predominate but certain advanced microwave and high-performance versions now employ the *compound semiconductor* material gallium arsenide (GaAs) and the *semiconductor alloy* silicon germanium (SiGe). Single element semiconductor material (Ge and Si) is described as *elemental*.

Rough parameters for the most common semiconductor materials used to make transistors are given in the adjacent table; these parameters will vary with increase in temperature, electric field, impurity level, strain, and sundry other factors.

The *junction forward voltage* is the voltage applied to the emitter–base junction of a BJT in order to make the base conduct a specified current. The current increases exponentially as the junction forward voltage is increased. The values given in the table are typical for a current of 1 mA (the same values apply to semiconductor diodes). The lower the junction forward voltage the better, as this means that less power is required to "drive" the transistor. The junction forward voltage for a given current decreases with increase in temperature. For a typical silicon junction the change is -2.1 mV/°C. In some circuits special compensating elements (sensistors) must be used to compensate for such changes.

The density of mobile carriers in the channel of a MOSFET is a function of the electric field forming the channel and of various other phenomena such as the impurity level in the channel. Some impurities, called dopants, are introduced deliberately in making a MOSFET, to control the MOSFET electrical behavior.

The *electron mobility* and *hole mobility* columns show the average speed that electrons and holes diffuse through the semiconductor material with an electric field of 1 volt per meter applied across the material. In general, the higher the electron mobility the faster the transistor can operate. The table indicates that Ge is a better material than Si in this respect. However, Ge has four major shortcomings compared to silicon and gallium arsenide:

- Its maximum temperature is limited;

- it has relatively high leakage current;

- it cannot withstand high voltages;

- it is less suitable for fabricating integrated circuits.

Because the electron mobility is higher than the hole mobility for all semiconductor materials, a given bipolar n–p–n transistor tends to be swifter than an equivalent p–n–p transistor. GaAs has the highest electron mobility of the three semiconductors. It is for this reason that GaAs is used in high-frequency applications. A relatively recent FET development, the *high-electron-mobility transistor* (HEMT), has a heterostructure (junction between different semiconductor materials) of aluminium gallium arsenide (AlGaAs)-gallium arsenide (GaAs) which has twice the electron mobility of a GaAs-metal barrier junction. Because of their high speed and low noise, HEMTs are used in satellite receivers working at frequencies around 12 GHz. HEMTs based on gallium nitride and aluminium gallium nitride (AlGaN/GaN HEMTs) provide a still higher electron mobility and are being developed for various applications.

Max. junction temperature values represent a cross section taken from various manufacturers' data sheets. This temperature should not be exceeded or the transistor may be damaged.

Al–Si junction refers to the high-speed (aluminum–silicon) metal–semiconductor barrier diode, commonly known as a Schottky diode. This is included in the table because some silicon power IGFETs have a *parasitic* reverse Schottky diode formed between the source and drain as part of the fabrication process. This diode can be a nuisance, but sometimes it is used in the circuit.

Packaging

Assorted discrete transistors.

Soviet KT315b transistors.

Discrete transistors are individually packaged transistors. Transistors come in many different semiconductor packages. The two main categories are *through-hole* (or *lead-ed*), and *surface-mount*, also known as *surface-mount device* (SMD). The *ball grid array* (BGA) is the latest surface-mount package (currently only for large integrated circuits). It has solder "balls" on the underside in place of leads. Because they are smaller and have shorter interconnections, SMDs have better high-frequency characteristics but lower power rating.

Transistor packages are made of glass, metal, ceramic, or plastic. The package often dictates the power rating and frequency characteristics. Power transistors have larger packages that can be clamped to heat sinks for enhanced cooling. Additionally, most power transistors have the collector or drain physically connected to the metal enclosure. At the other extreme, some surface-mount *microwave* transistors are as small as grains of sand.

Often a given transistor type is available in several packages. Transistor packages are mainly standardized, but the assignment of a transistor's functions to the terminals is not: other transistor types can assign other functions to the package's terminals. Even for the same transistor type the terminal assignment can vary (normally indicated by a suffix letter to the part number, q.e. BC212L and BC212K).

Nowadays most transistors come in a wide range of SMT packages, in comparison the list of available through-hole packages is relatively small, here is a short list of the most common through-hole transistors packages in alphabetical order: ATV, E-line, MRT, HRT, SC-43, SC-72, TO-3, TO-18, TO-39, TO-92, TO-126, TO220, TO247, TO251, TO262, ZTX851.

Flexible Transistors

Researchers have made several kinds of flexible transistors, including organic field-effect transistors. Flexible transistors are useful in some kinds of flexible displays and other flexible electronics.

Bipolar Junction Transistor

A transistor is basically a Si on Ge crystal containing three separate regions. It can be either NPN or PNP type in the figure below. The middle region is called the base and the outer two regions are called emitter and the collector. The outer layers although they are of same type but their functions cannot be changed. They have different physical and electrical properties.

In most transistors, emitter is heavily doped. Its job is to emit or inject electrons into the base. These bases are lightly doped and very thin, it passes most of the emitter-injected electrons on to the collector. The doping level of collector is intermediate between the heavy doping of emitter and the light doping of the base.

The collector is so named because it collects electrons from base. The collector is the largest of the three regions; it must dissipate more heat than the emitter or base. The transistor has two junctions. One between emitter and the base and other between the base and the collector. Because of this the transistor is similar to two diodes, one emitter diode and other collector base diode.

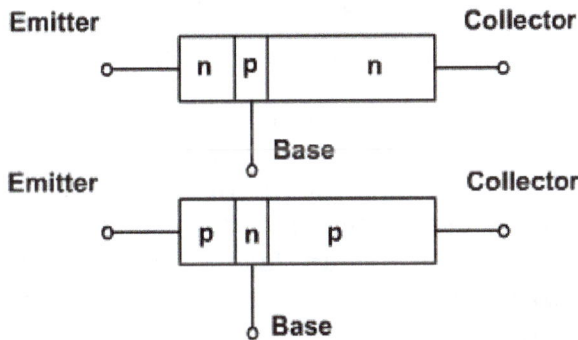

When transistor is made, the diffusion of free electrons across the junction produces two depletion layers. For each of these depletion layers, the barrier potential is 0.7 V for Si transistor and 0.3 V for Ge transistor.

The depletion layers do not have the same width, because different regions have different doping levels. The more heavily doped a region is, the greater the concentration of ions near the junction. This means the depletion layer penetrates more deeply into the base and slightly into emitter. Similarly, it penetration more into collector. The thickness of collector depletion layer is large while the base depletion layer is small as shown in the below figure.

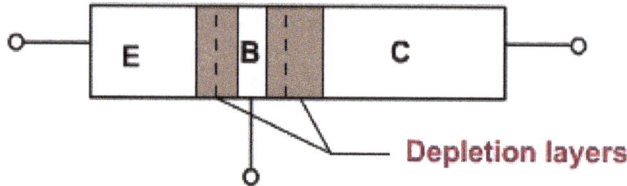

Depletion layers

If both the junctions are forward biased using two d.c sources, as shown in the fig. a. free electrons (majority carriers) enter the emitter and collector of the transistor, joins at the base and come out of the base. Because both the diodes are forward biased, the emitter and collector currents are large.

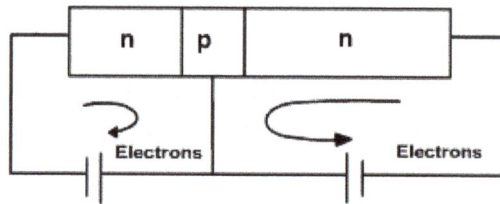

Fig. a

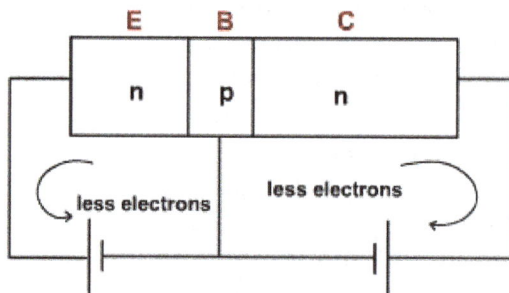

Fig. b

If both the junction are reverse biased as shown in fig. b, then small currents flows through both junctions only due to thermally produced minority carriers and surface leakage. Thermally produced carriers are temperature dependent it approximately doubles for every 10 degree celsius rise in ambient temperature. The surface leakage current increases with voltage.

When the emitter diode is forward biased and collector diode is reverse biased as shown in the figure given below. then one expect large emitter current and small collector current but collector current is almost as large as emitter current.

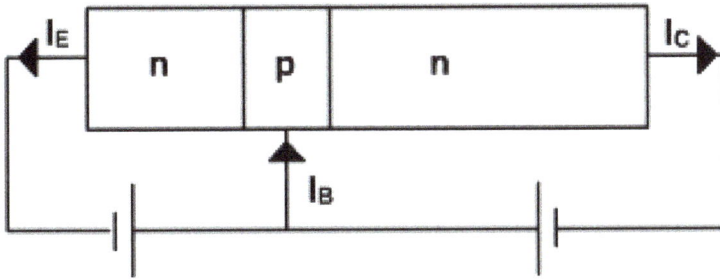

When emitter diodes forward biased and the applied voltage is more than 0.7 V (barrier potential) then larger number of majority carriers (electrons in n-type) diffuse across the junction.

Once the electrons are injected by the emitter enter into the base, they become minority carriers. These electrons do not have separate identities from those, which are thermally generated, in the base region itself. The base is made very thin and is very lightly doped. Because of this only few electrons traveling from the emitter to base region recombine with holes. This gives rise to recombination current. The rest of the electrons exist for more time. Since the collector diode is reverse biased, (n is connected to positive supply) therefore most of the electrons are pushed into collector layer. These collector elections can then flow into the external collector lead.

Thus, there is a steady stream of electrons leaving the negative source terminal and entering the emitter region. The V_{EB} forward bias forces these emitter electrons to enter the base region. The thin and lightly doped base gives almost all those electrons enough lifetime to diffuse into the depletion layer. The depletion layer field pushes a steady stream of electron into the collector region. These electrons leave the collector and flow into the positive terminal of the voltage source. In most transistor, more than 95% of the emitter injected electrons flow to the collector, less than 5% fall into base holes and flow out the external base lead. But the collector current is less than emitter current.

Relation Between Different Currents in a Transistor

The total current flowing into the transistor must be equal to the total current flowing out of it. Hence, the emitter current I_E is equal to the sum of the collector (I_C) and base current (I_B). That is,

$$I_E = I_C + I_B$$

The currents directions are positive directions. The total collector current I_C is made up of two components.

1. The fraction of emitter (electron) current which reaches the collector ($\alpha_{dc} I_E$)

2. The normal reverse leakage current I_{CO}

$$\therefore I_C = \alpha_{dc} I_E + I_\infty$$

$$\text{or} \quad \alpha_{dc} = \frac{I_C - I_\infty}{I_E}$$

α_{dc} is known as large signal current gain or dc alpha. It is always positive. Since collector current is almost equal to the I_E therefore $\alpha_{dc} I_E$ varies from 0.9 to 0.98. Usually, the reverse leakage current is very small compared to the total collector current.

$$\text{Neglecting } I_{CO}, \quad \alpha_{dc} = \frac{I_C}{I_E}$$

NOTE: The forward bias on the emitter diode controls the number of free electrons infected into the base. The larger (V_{BE}) forward voltage, the greater the number of injected electrons. The reverse bias on the collector diode has little influence on the number of electrons that enter the collector. Increasing V_{CB} does not change the number of free electrons arriving at the collector junction layer.

The symbol of npn and pnp transistors are shown in the figure below.

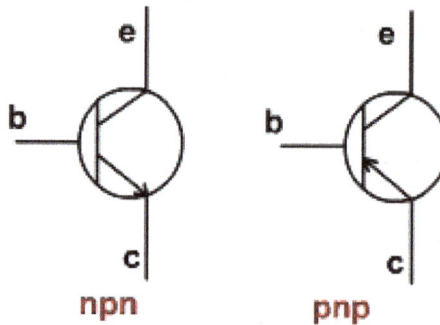

npn   pnp

Breakdown Voltages

Since the two halves of a transistor are diodes, two much reverse voltage on either diode can cause breakdown. The breakdown voltage depends on the width of the depletion layer and the doping levels. Because of the heavy doping level, the emitter diode has a low breakdown voltage approximately 5 to 30 V. The collector diode is less heavily doped so its breakdown voltage is higher around 20 to 300 V.

A bipolar junction transistor (bipolar transistor or BJT) is a type of transistor that uses both electron and hole charge carriers. In contrast, unipolar transistors, such as field-effect transistors, only use one kind of charge carrier. For their operation, BJTs use two junctions between two semiconductor types, n-type and p-type.

BJTs are manufactured in two types, NPN and PNP, and are available as individual components, or fabricated in integrated circuits, often in large numbers. The basic function of a BJT is to amplify current. This allows BJTs to be used as amplifiers or switches, giving them wide applicability in electronic equipment, including computers, televisions, mobile phones, audio amplifiers, industrial control, and radio transmitters.

Note on Current Direction

By convention, the direction of current on diagrams is shown as the direction that a positive charge would move. This is called *conventional current*. However, current in many metal conductors is due to the flow of electrons which, because they carry a negative charge, move in the opposite direction to conventional current. On the other hand, inside a bipolar transistor, currents can be composed of both positively charged holes and negatively charged electrons. Current arrows are shown in the conventional direction, but labels for the movement of holes and electrons show their actual direction inside the transistor. The arrow on the symbol for bipolar transistors points in the direction conventional current travels.

Function

BJTs come in two types, or polarities, known as PNP and NPN based on the doping types of the three main terminal regions. An NPN transistor comprises two semiconductor junctions that share a thin p-doped region, and a PNP transistor comprises two semiconductor junctions that share a thin n-doped region.

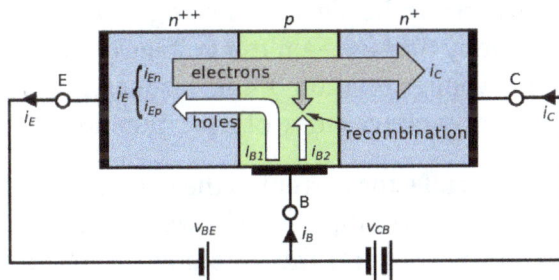

NPN BJT with forward-biased E–B junction and reverse-biased B–C junction

Charge flow in a BJT is due to diffusion of charge carriers across a junction between two regions of different charge concentrations. The regions of a BJT are called *emitter*, *collector*, and *base*. A discrete transistor has three leads for connection to these regions. Typically, the emitter region is heavily doped compared to the other two layers, whereas the majority charge carrier concentrations in base and collector layers are about the same. By design, most of the BJT collector current is due to the flow of charges injected from a high-concentration emitter into the base where they are minority carriers that diffuse toward the collector, and so BJTs are classified as minority-carrier devices.

In typical operation, the base–emitter junction is forward biased, which means that the p-doped side of the junction is at a more positive potential than the n-doped side, and the base–collector junction is reverse biased. In an NPN transistor, when positive bias is applied to the base–emitter junction, the equilibrium is disturbed between the thermally generated carriers and the repelling electric field of the n-doped emitter depletion region. This allows thermally excited electrons to inject from the emitter into the base region. These electrons diffuse through the base from the region of high concentration near the emitter towards the region of low concentration near the collector. The electrons in the base are called *minority carriers* because the base is doped p-type, which makes holes the *majority carrier* in the base.

To minimize the percentage of carriers that recombine before reaching the collector–base junction, the transistor's base region must be thin enough that carriers can diffuse across it in much less time than the semiconductor's minority carrier lifetime. In particular, the thickness of the base must be much less than the diffusion length of the electrons. The collector–base junction is reverse-biased, and so little electron injection occurs from the collector to the base, but electrons that diffuse through the base towards the collector are swept into the collector by the electric field in the depletion region of the collector–base junction. The thin *shared* base and asymmetric collector–emitter doping are what differentiates a bipolar transistor from two *separate* and oppositely biased diodes connected in series.

Voltage, Current, and Charge Control

The collector–emitter current can be viewed as being controlled by the base–emitter current (current control), or by the base–emitter voltage (voltage control). These views are related by the current–voltage relation of the base–emitter junction, which is just the usual exponential current–voltage curve of a p-n junction (diode).

The physical explanation for collector current is the concentration of minority carriers in the base region. Due to low level injection (in which there are much fewer excess carriers than normal majority carriers) the ambipolar transport rates (in which the excess majority and minority carriers flow at the same rate) is in effect determined by the excess minority carriers.

Detailed transistor models of transistor action, such as the Gummel–Poon model, account for the distribution of this charge explicitly to explain transistor behaviour more exactly. The charge-control view easily handles phototransistors, where minority carriers in the base region are created by the absorption of photons, and handles the dynamics of turn-off, or recovery time, which depends on charge in the base region recombining. However, because base charge is not a signal that is visible at the terminals, the current- and voltage-control views are generally used in circuit design and analysis.

In analog circuit design, the current-control view is sometimes used because it is approximately linear. That is, the collector current is approximately β_F times the base current. Some basic circuits can be designed by assuming that the emitter–base voltage is approximately constant, and that collector current is beta times the base current. However, to accurately and reliably design production BJT circuits, the voltage-control (for example, Ebers–Moll) model is required. The voltage-control model requires an exponential function to be taken into account, but when it is linearized such that the transistor can be modeled as a transconductance, as in the Ebers–Moll model, design for circuits such as differential amplifiers again becomes a mostly linear problem, so the voltage-control view is often preferred. For translinear circuits, in which the exponential I–V curve is key to the operation, the transistors are usually modeled as voltage-controlled current sources whose transconductance is proportional to their collector current. In general, transistor-level circuit design is performed using SPICE or a comparable analog circuit simulator, so model complexity is usually not of much concern to the designer.

Turn-on, Turn-off, and Storage Delay

The bipolar transistor exhibits a few delay characteristics when turning on and off. Most transistors, and especially power transistors, exhibit long base-storage times that limit maximum frequency of operation in switching applications. One method for reducing this storage time is by using a Baker clamp.

Transistor Parameters: Alpha (α) and Beta (β)

The proportion of electrons able to cross the base and reach the collector is a measure of the BJT efficiency. The heavy doping of the emitter region and light doping of the base region causes many more electrons to be injected from the emitter into the base than holes to be injected from the base into the emitter.

The *common-emitter current gain* is represented by β_F or the h-parameter h_{FE}; it is approximately the ratio of the DC collector current to the DC base current in forward-active region. It is typically greater than 50 for small-signal transistors but can be smaller in transistors designed for high-power applications.

Another important parameter is the *common-base current gain*, α_F. The common-base current gain is approximately the gain of current from emitter to collector in the forward-active region. This ratio usually has a value close to unity; between 0.980 and 0.998. It is less than unity due to recombination of charge carriers as they cross the base region.

Alpha and beta are more precisely related by the following identities (NPN transistor):

$$\alpha_F = \frac{I_C}{I_E}, \qquad \beta_F = \frac{I_C}{I_B}$$

$$\alpha_F = \frac{\beta_F}{1+\beta_F} \Leftrightarrow \beta_F = \frac{\alpha_F}{1-\alpha_F}$$

Structure

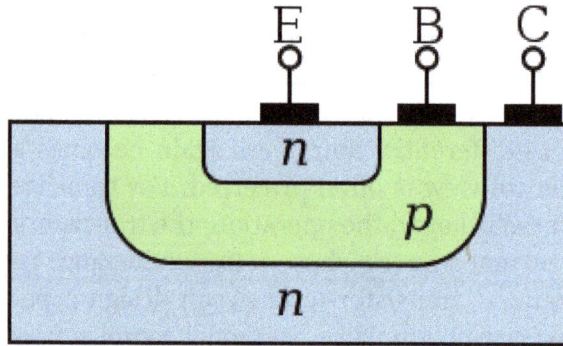

Simplified cross section of a planar *NPN* bipolar junction transistor

A BJT consists of three differently doped semiconductor regions: the *emitter* region, the *base* region and the *collector* region. These regions are, respectively, *p* type, *n* type and *p* type in a PNP transistor, and *n* type, *p* type and *n* type in an NPN transistor. Each semiconductor region is connected to a terminal, appropriately labeled: *emitter* (E), *base* (B) and *collector* (C).

The *base* is physically located between the *emitter* and the *collector* and is made from lightly doped, high-resistivity material. The collector surrounds the emitter region, making it almost impossible for the electrons injected into the base region to escape without being collected, thus making the resulting value of α very close to unity, and so, giving the transistor a large β. A cross-section view of a BJT indicates that the collector–base junction has a much larger area than the emitter–base junction.

Die of a KSY34 high-frequency NPN transistor. Bond wires connect to the base and emitter

The bipolar junction transistor, unlike other transistors, is usually not a symmetrical device. This means that interchanging the collector and the emitter makes the transistor leave the forward active mode and start to operate in reverse mode. Because the transistor's internal structure is usually optimized for forward-mode operation, interchanging the collector and the emitter makes the values of α and β in reverse operation much smaller than those in forward operation; often the α of the reverse mode is lower than 0.5. The lack of symmetry is primarily due to the doping ratios of the emitter and the collector. The emitter is heavily doped, while the collector is lightly doped, allowing a large reverse bias voltage to be applied before the collector–base junction breaks down. The collector–base junction is reverse biased in normal operation. The reason the emitter is heavily doped is to increase the emitter injection efficiency: the ratio of carriers injected by the emitter to those injected by the base. For high current gain, most of the carriers injected into the emitter–base junction must come from the emitter.

The low-performance "lateral" bipolar transistors sometimes used in CMOS processes are sometimes designed symmetrically, that is, with no difference between forward and backward operation.

Small changes in the voltage applied across the base–emitter terminals cause the current between the *emitter* and the *collector* to change significantly. This effect can be used to amplify the input voltage or current. BJTs can be thought of as voltage-controlled current sources, but are more simply characterized as current-controlled current sources, or current amplifiers, due to the low impedance at the base.

Early transistors were made from germanium but most modern BJTs are made from silicon. A significant minority are also now made from gallium arsenide, especially for very high speed applications.

NPN

The symbol of an NPN BJT. A mnemonic for the symbol is "not pointing in".

NPN is one of the two types of bipolar transistors, consisting of a layer of P-doped semiconductor (the "base") between two N-doped layers. A small current entering the base is amplified to produce a large collector and emitter current. That is, when there is a positive potential difference measured from the base of an NPN transistor to its emitter (that is, when the base is high relative to the emitter), as well as a positive potential difference measured from the collector to the emitter, the transistor becomes active. In this "on" state, charge flows from the collector to the emitter of the transistor. Most of the current is carried by electrons moving from emitter to collector as minority carriers in the P-type base region. To allow for greater current and faster operation, most bipolar transistors used today are NPN because electron mobility is higher than hole mobility.

A mnemonic device for the NPN transistor symbol is *"not pointing in"*, based on the arrows in the symbol and the letters in the name.

PNP

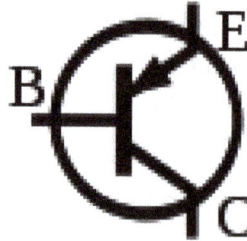

The symbol of a PNP BJT. A mnemonic for the symbol is "points in proudly".

The other type of BJT is the PNP, consisting of a layer of N-doped semiconductor between two layers of P-doped material. A small current leaving the base is amplified in the collector output. That is, a PNP transistor is "on" when its base is pulled low relative to the emitter. In a PNP transistor, the emitter–base region is forward biased, so holes are injected into the base as minority carriers. The base is very thin, and most of the holes cross the reverse-biased base–collector junction to the collector.

The arrows in the NPN and PNP transistor symbols are on the emitter legs and point in the direction of the conventional current when the device is in forward active or forward saturated mode.

A mnemonic device for the PNP transistor symbol is *"pointing in (proudly/permanently)"*, based on the arrows in the symbol and the letters in the name.

Heterojunction Bipolar Transistor

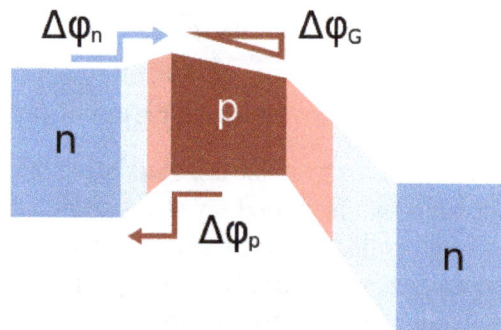

Bands in graded heterojunction NPN bipolar transistor. Barriers indicated for electrons to move from emitter to base, and for holes to be injected backward from base to emitter; also, grading of bandgap in base assists electron transport in base region; light colors indicate depleted regions.

The heterojunction bipolar transistor (HBT) is an improvement of the BJT that can handle signals of very high frequencies up to several hundred GHz. It is common in modern ultrafast circuits, mostly RF systems.

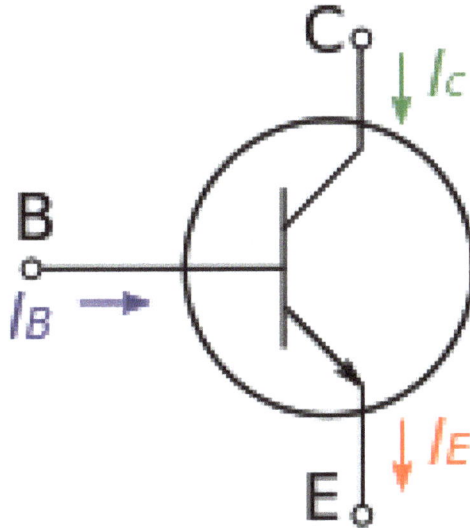

Symbol for NPN Bipolar Transistor with current flow direction.

Heterojunction transistors have different semiconductors for the elements of the transistor. Usually the emitter is composed of a larger bandgap material than the base. The figure shows that this difference in bandgap allows the barrier for holes to inject backward from the base into the emitter, denoted in the figure as $\Delta\varphi_p$, to be made large, while the barrier for electrons to inject into the base $\Delta\varphi_n$ is made low. This barrier arrangement helps reduce minority carrier injection from the base when the emitter-base junction is under forward bias, and thus reduces base current and increases emitter injection efficiency.

The improved injection of carriers into the base allows the base to have a higher doping level, resulting in lower resistance to access the base electrode. In the more traditional BJT, also referred to as homojunction BJT, the efficiency of carrier injection from the emitter to the base is primarily determined by the doping ratio between the emitter and base, which means the base must be lightly doped to obtain high injection efficiency, making its resistance relatively high. In addition, higher doping in the base can improve figures of merit like the Early voltage by lessening base narrowing.

The grading of composition in the base, for example, by progressively increasing the amount of germanium in a SiGe transistor, causes a gradient in bandgap in the neutral base, denoted in the figure by $\Delta\varphi_G$, providing a "built-in" field that assists electron transport across the base. That drift component of transport aids the normal diffusive transport, increasing the frequency response of the transistor by shortening the transit time across the base.

Two commonly used HBTs are silicon–germanium and aluminum gallium arsenide, though a wide variety of semiconductors may be used for the HBT structure. HBT structures are usually grown by epitaxy techniques like MOCVD and MBE.

Regions of Operation

Applied voltages	B-E junction bias (NPN)	B-C junction bias (NPN)	Mode (NPN)
E < B < C	Forward	Reverse	Forward-active
E < B > C	Forward	Forward	Saturation
E > B < C	Reverse	Reverse	Cut-off
E > B > C	Reverse	Forward	Reverse-active
Applied voltages	B-E junction bias (PNP)	B-C junction bias (PNP)	Mode (PNP)
E < B < C	Reverse	Forward	Reverse-active
E < B > C	Reverse	Reverse	Cut-off
E > B < C	Forward	Forward	Saturation
E > B > C	Forward	Reverse	Forward-active

Bipolar transistors have four distinct regions of operation, defined by BJT junction biases.

Forward-active (or Simply *Active*)

> The base–emitter junction is forward biased and the base–collector junction is reverse biased. Most bipolar transistors are designed to afford the greatest common-emitter current gain, β_F, in forward-active mode. If this is the case, the collector–emitter current is approximately proportional to the base current, but many times larger, for small base current variations.

Reverse-active (or *inverse-active* or *inverted*)

> By reversing the biasing conditions of the forward-active region, a bipolar transistor goes into reverse-active mode. In this mode, the emitter and collector regions switch roles. Because most BJTs are designed to maximize current gain in forward-active mode, the β_F in inverted mode is several times smaller (2–3 times for the ordinary germanium transistor). This transistor mode is seldom used, usually being considered only for failsafe conditions and some types of bipolar logic. The reverse bias breakdown voltage to the base may be an order of magnitude lower in this region.

Saturation

> With both junctions forward-biased, a BJT is in saturation mode and facilitates high current conduction from the emitter to the collector (or the other direction in the case of NPN, with negatively charged carriers flowing from emitter to collector). This mode corresponds to a logical "on", or a closed switch.

Cut-off

> In cut-off, biasing conditions opposite of saturation (both junctions reverse bi-

ased) are present. There is very little current, which corresponds to a logical "off", or an open switch.

Avalanche breakdown region

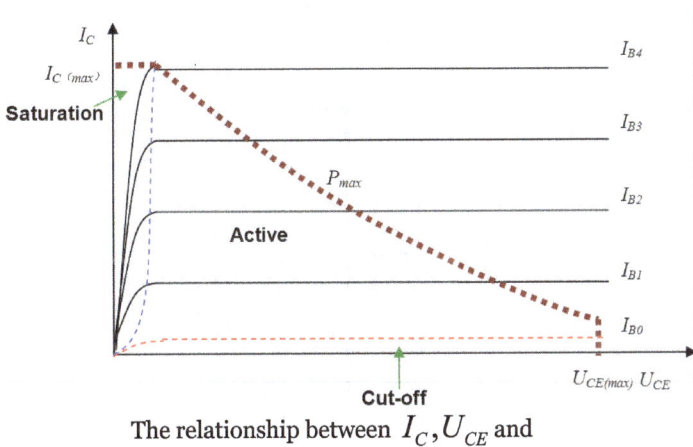

The relationship between I_C, U_{CE} and

The modes of operation can be described in terms of the applied voltages (this description applies to NPN transistors; polarities are reversed for PNP transistors):

Forward-active

Base higher than emitter, collector higher than base (in this mode the collector current is proportional to base current by β_F).

Saturation

Base higher than emitter, but collector is not higher than base.

Cut-off

Base lower than emitter, but collector is higher than base. It means the transistor is not letting conventional current go through from collector to emitter.

Reverse-active

Base lower than emitter, collector lower than base: reverse conventional current goes through transistor.

In terms of junction biasing: (*reverse biased base–collector junction* means $V_{bc} < 0$ for NPN, opposite for PNP)

Although these regions are well defined for sufficiently large applied voltage, they overlap somewhat for small (less than a few hundred millivolts) biases. For example, in the typical grounded-emitter configuration of an NPN BJT used as a pulldown switch in digital logic, the "off" state never involves a reverse-biased junction because the base voltage never goes below ground; nevertheless the forward bias is close enough to zero that essentially no current flows, so this end of the forward active region can be regarded as the cutoff region.

Active-mode NPN Transistors in Circuits

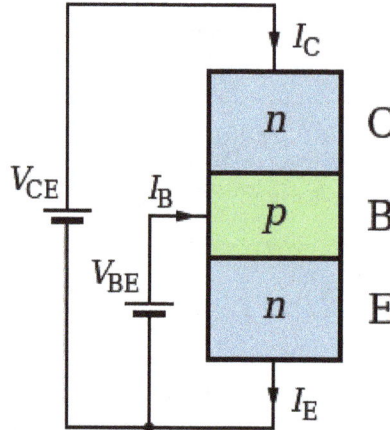

Structure and use of NPN transistor. Arrow according to schematic.

The diagram shows a schematic representation of an NPN transistor connected to two voltage sources. To make the transistor conduct appreciable current (on the order of 1 mA) from C to E, V_{BE} must be above a minimum value sometimes referred to as the cut-in voltage. The cut-in voltage is usually about 650 mV for silicon BJTs at room temperature but can be different depending on the type of transistor and its biasing. This applied voltage causes the lower P-N junction to 'turn on', allowing a flow of electrons from the emitter into the base. In active mode, the electric field existing between base and collector (caused by V_{CE}) will cause the majority of these electrons to cross the upper P-N junction into the collector to form the collector current I_C. The remainder of the electrons recombine with holes, the majority carriers in the base, making a current through the base connection to form the base current, I_B. As shown in the diagram, the emitter current, I_E, is the total transistor current, which is the sum of the other terminal currents, (i.e., $I_E = I_B + I_C$).

In the diagram, the arrows representing current point in the direction of conventional current – the flow of electrons is in the opposite direction of the arrows because electrons carry negative electric charge. In active mode, the ratio of the collector current to the base current is called the *DC current gain*. This gain is usually 100 or more, but robust circuit designs do not depend on the exact value (for example, op-amp). The value of this gain for DC signals is referred to as h_{FE}, and the value of this gain for small signals is referred to as h_{fe}. That is, when a small change in the currents occurs, and sufficient time has passed for the new condition to reach a steady state h_{fe} is the ratio of the change in collector current to the change in base current. The symbol β is used for both h_{FE} and h_{fe}.

The emitter current is related to V_{BE} exponentially. At room temperature, an increase in V_{BE} by approximately 60 mV increases the emitter current by a factor of 10. Because the base current is approximately proportional to the collector and emitter currents, they vary in the same way.

Active-mode PNP Transistors in Circuits

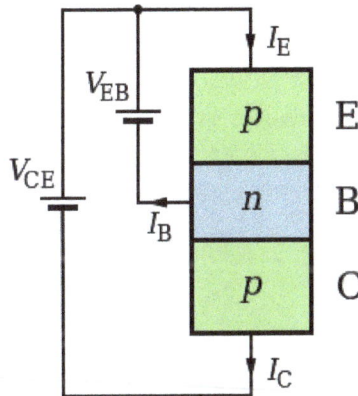

Structure and use of PNP transistor

The diagram shows a schematic representation of a PNP transistor connected to two voltage sources. To make the transistor conduct appreciable current (on the order of 1 mA) from E to C, V_{EB} must be above a minimum value sometimes referred to as the cut-in voltage. The cut-in voltage is usually about 650 mV for silicon BJTs at room temperature but can be different depending on the type of transistor and its biasing. This applied voltage causes the upper P-N junction to 'turn-on' allowing a flow of holes from the emitter into the base. In active mode, the electric field existing between the emitter and the collector (caused by V_{CE}) causes the majority of these holes to cross the lower p-n junction into the collector to form the collector current I_C. The remainder of the holes recombine with electrons, the majority carriers in the base, making a current through the base connection to form the base current, I_B. As shown in the diagram, the emitter current, I_E, is the total transistor current, which is the sum of the other terminal currents (i.e., $I_E = I_B + I_C$).

In the diagram, the arrows representing current point in the direction of conventional current – the flow of holes is in the same direction of the arrows because holes carry positive electric charge. In active mode, the ratio of the collector current to the base current is called the *DC current gain*. This gain is usually 100 or more, but robust circuit designs do not depend on the exact value. The value of this gain for DC signals is referred to as h_{FE}, and the value of this gain for AC signals is referred to as h_{fe}. However, when there is no particular frequency range of interest, the symbol β is used.

The emitter current is related to V_{EB} exponentially. At room temperature, an increase in V_{EB} by approximately 60 mV increases the emitter current by a factor of 10. Because the base current is approximately proportional to the collector and emitter currents, they vary in the same way.

History

The bipolar point-contact transistor was invented in December 1947 at the Bell Telephone Laboratories by John Bardeen and Walter Brattain under the direction of William Shockley. The junction version known as the bipolar junction transistor (BJT), invented by Shockley in 1948, was for three decades the device of choice in the design of discrete and integrated circuits. Nowadays, the use of the BJT has declined in favor of CMOS technology in the design of digital integrated circuits. The incidental low performance BJTs inherent in CMOS ICs, however, are often utilized as bandgap voltage reference, silicon bandgap temperature sensor and to handle electrostatic discharge.

Germanium Transistors

The germanium transistor was more common in the 1950s and 1960s, and while it exhibits a lower "cut-off" voltage, typically around 0.2 V, making it more suitable for some applications, it also has a greater tendency to exhibit thermal runaway.

Early Manufacturing Techniques

Various methods of manufacturing bipolar transistors were developed.

Bipolar Transistors

- Point-contact transistor – first transistor ever constructed (December 1947), a bipolar transistor, limited commercial use due to high cost and noise.

  o Tetrode point-contact transistor – Point-contact transistor having two emitters. It became obsolete in the middle 1950s.

- Junction transistors

  o Grown-junction transistor – first bipolar *junction* transistor made. Invented by William Shockley at Bell Labs on June 23, 1948. Patent filed on June 26, 1948.

  o Alloy-junction transistor – emitter and collector alloy beads fused to base. Developed at General Electric and RCA in 1951.

- Micro-alloy transistor (MAT) – high speed type of alloy junction transistor. Developed at Philco.

- Micro-alloy diffused transistor (MADT) – high speed type of alloy junction transistor, speedier than MAT, a diffused-base transistor. Developed at Philco.

- Post-alloy diffused transistor (PADT) – high speed type of alloy junction transistor, speedier than MAT, a diffused-base transistor. Developed at Philips.

  o Tetrode transistor – high speed variant of grown-junction transistor or alloy junction transistor with two connections to base.

  o Surface-barrier transistor – high-speed metal barrier junction transistor. Developed at Philco in 1953.

  o Drift-field transistor – high speed bipolar junction transistor. Invented by Herbert Kroemer at the Central Bureau of Telecommunications Technology of the German Postal Service, in 1953.

  o Spacistor – circa 1957.

  o Diffusion transistor – modern type bipolar junction transistor. Prototypes developed at Bell Labs in 1954.

- Diffused-base transistor – first implementation of diffusion transistor.

- Mesa transistor – Developed at Texas Instruments in 1957.

- Planar transistor – the bipolar junction transistor that made mass-produced monolithic integrated circuits possible. Developed by Jean Hoerni at Fairchild in 1959.

  o Epitaxial transistor – a bipolar junction transistor made using vapor phase deposition. Allows very precise control of doping levels and gradients.

Theory and Modeling

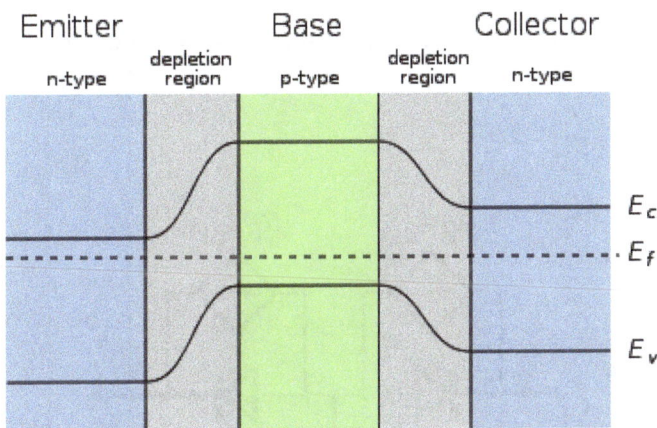

Band diagram for NPN transistor at equilibrium

Transistors can be thought of as two diodes (P–N junctions) sharing a common region that minority carriers can move through. A PNP BJT will function like two diodes that share an N-type cathode region, and the NPN like two diodes sharing a P-type anode region. Connecting two diodes with wires will not make a transistor, since minority carriers will not be able to get from one P–N junction to the other through the wire.

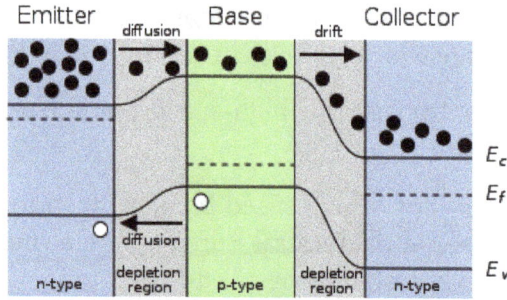

Band diagram for NPN transistor in active mode, showing injection of electrons from emitter to base, and their overshoot into the collector

Both types of BJT function by letting a small current input to the base control an amplified output from the collector. The result is that the transistor makes a good switch that is controlled by its base input. The BJT also makes a good amplifier, since it can multiply a weak input signal to about 100 times its original strength. Networks of transistors are used to make powerful amplifiers with many different applications. In the discussion below, focus is on the NPN bipolar transistor. In the NPN transistor in what is called active mode, the base–emitter voltage V_{BE} and collector–base voltage V_{CB} are positive, forward biasing the emitter–base junction and reverse-biasing the collector–base junction. In the active mode of operation, electrons are injected from the forward biased n-type emitter region into the p-type base where they diffuse as minority carriers to the reverse-biased n-type collector and are swept away by the electric field in the reverse-biased collector–base junction.

Large-signal Models

In 1954, Jewell James Ebers and John L. Moll introduced their mathematical model of transistor currents.

Ebers–Moll Model

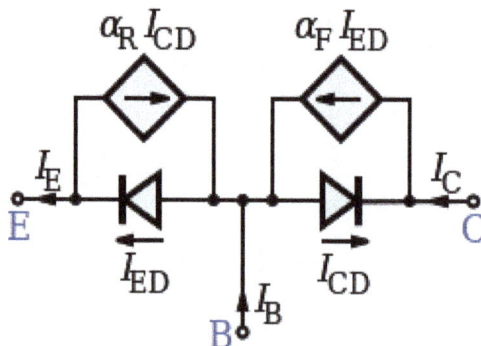

Ebers–Moll model for an NPN transistor * I_B, I_C, I_E: base, collector and emitter currents * I_{CD}, I_{ED}: collector and emitter diode currents * α_F, α_R: forward and reverse common-base current gains

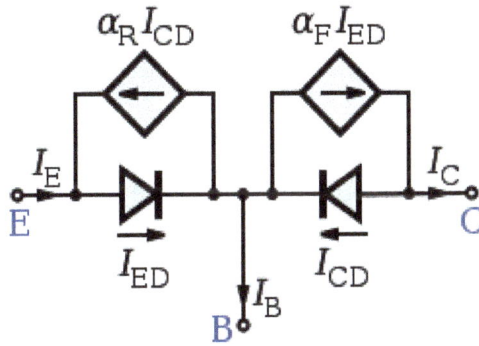

Ebers–Moll model for a PNP transistor

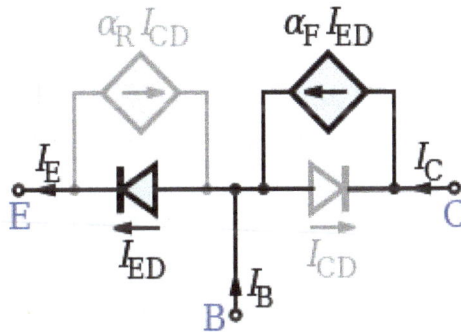

Approximated Ebers–Moll model for an NPN transistor in the forward active mode. The collector diode is reverse-biased so I_{CD} is virtually zero. Most of the emitter diode current (α_F is nearly 1) is drawn from the collector, providing the amplification of the base current.

The DC emitter and collector currents in active mode are well modeled by an approximation to the Ebers–Moll model:

$$I_E = I_{ES}\left(e^{\frac{V_{BE}}{V_T}} - 1\right)$$

$$I_C = \alpha_F I_E$$

$$I_B = (1 - \alpha_F) I_E$$

The base internal current is mainly by diffusion and

$$J_{n(base)} = \frac{1}{W} q D_n n_{bo} e^{\frac{V_{EB}}{V_T}}$$

where

- V_T is the thermal voltage kT / q (approximately 26 mV at 300 K ≈ room temperature).

- I_E is the emitter current

- I_C is the collector current
- α_F is the common base forward short-circuit current gain (0.98 to 0.998)
- I_{ES} is the reverse saturation current of the base–emitter diode (on the order of 10^{-15} to 10^{-12} amperes)
- V_{BE} is the base–emitter voltage
- D_n is the diffusion constant for electrons in the p-type base
- W is the base width

The α and forward β parameters are as described previously. A reverse β is sometimes included in the model.

The unapproximated Ebers–Moll equations used to describe the three currents in any operating region are given below. These equations are based on the transport model for a bipolar junction transistor.

$$i_C = I_S \left[\left(e^{\frac{V_{BE}}{V_T}} - e^{\frac{V_{BC}}{V_T}} \right) - \frac{1}{\beta_R} \left(e^{\frac{V_{BC}}{V_T}} - 1 \right) \right]$$

$$i_B = I_S \left[\frac{1}{\beta_F} \left(e^{\frac{V_{BE}}{V_T}} - 1 \right) + \frac{1}{\beta_R} \left(e^{\frac{V_{BC}}{V_T}} - 1 \right) \right]$$

$$i_E = I_S \left[\left(e^{\frac{V_{BE}}{V_T}} - e^{\frac{V_{BC}}{V_T}} \right) + \frac{1}{\beta_F} \left(e^{\frac{V_{BE}}{V_T}} - 1 \right) \right]$$

where

- i_C is the collector current
- i_B is the base current
- i_E is the emitter current
- β_F is the forward common emitter current gain (20 to 500)
- β_R is the reverse common emitter current gain (0 to 20)
- I_S is the reverse saturation current (on the order of 10^{-15} to 10^{-12} amperes)
- V_T is the thermal voltage (approximately 26 mV at 300 K ≈ room temperature).
- V_{BE} is the base–emitter voltage
- V_{BC} is the base–collector voltage

Base-width Modulation

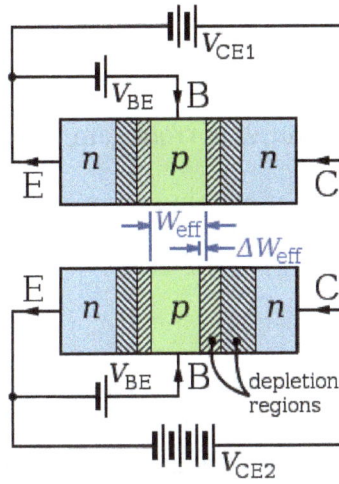

Top: NPN base width for low collector-base reverse bias; Bottom: narrower NPN base width for large collector-base reverse bias. Hashed regions are depleted regions.

As the collector–base voltage ($V_{CB} = V_{CE} - V_{BE}$) varies, the collector–base depletion region varies in size. An increase in the collector–base voltage, for example, causes a greater reverse bias across the collector–base junction, increasing the collector–base depletion region width, and decreasing the width of the base. This variation in base width often is called the "Early effect" after its discoverer James M. Early.

Narrowing of the base width has two consequences:

- There is a lesser chance for recombination within the "smaller" base region.

- The charge gradient is increased across the base, and consequently, the current of minority carriers injected across the emitter junction increases.

Both factors increase the collector or "output" current of the transistor in response to an increase in the collector–base voltage.

In the forward-active region, the Early effect modifies the collector current (i_C) and the forward common emitter current gain (β_F) as given by:

$$i_C = I_S e^{\frac{v_{BE}}{V_T}} \left(1 + \frac{V_{CE}}{V_A} \right)$$

$$\hat{a}_F = \hat{a}_{F0} \left(1 + \frac{V_{CB}}{V_A} \right)$$

$$r_o = \frac{V_A}{I_C}$$

where:

- V_{CE} is the collector–emitter voltage
- V_A is the Early voltage (15 V to 150 V)
- β_{F0} is forward common-emitter current gain when $V_{CB} = 0$ V
- r_o is the output impedance
- I_C is the collector current

Punchthrough

When the base–collector voltage reaches a certain (device-specific) value, the base–collector depletion region boundary meets the base–emitter depletion region boundary. When in this state the transistor effectively has no base. The device thus loses all gain when in this state.

Gummel–Poon Charge-control Model

The Gummel–Poon model is a detailed charge-controlled model of BJT dynamics, which has been adopted and elaborated by others to explain transistor dynamics in greater detail than the terminal-based models typically do. This model also includes the dependence of transistor β -values upon the direct current levels in the transistor, which are assumed current-independent in the Ebers–Moll model.

Small-signal Models

Hybrid-pi Model

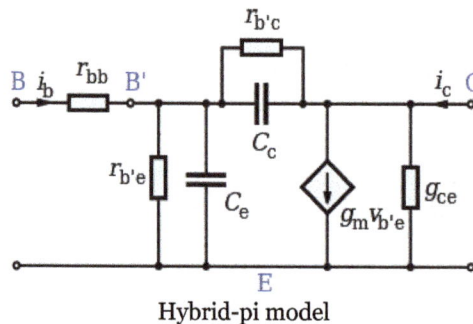

Hybrid-pi model

The hybrid-pi model is a popular circuit model used for analyzing the small signal behavior of bipolar junction and field effect transistors. Sometimes it is also called *Giacoletto model* because it was introduced by L.J. Giacoletto in 1969. The model can be quite accurate for low-frequency circuits and can easily be adapted for higher-frequency circuits with the addition of appropriate inter-electrode capacitances and other parasitic elements.

h-parameter Model

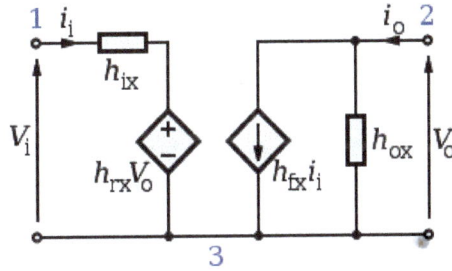

Generalized h-parameter model of an NPN BJT.
Replace x with e, b or c for CE, CB and CC topologies respectively.

Another model commonly used to analyze BJT circuits is the *h-parameter* model, close-
ly related to the hybrid-pi model and the y-parameter two-port, but using input current
and output voltage as independent variables, rather than input and output voltages.
This two-port network is particularly suited to BJTs as it lends itself easily to the analysis of circuit behaviour, and may be used to develop further accurate models. As shown,
the term, x, in the model represents a different BJT lead depending on the topology
used. For common-emitter mode the various symbols take on the specific values as:

- Terminal 1, base

- Terminal 2, collector

- Terminal 3 (common), emitter; giving x to be e

- i_i, base current (i_b)

- i_o, collector current (i_c)

- V_{in}, base-to-emitter voltage (V_{BE})

- V_o, collector-to-emitter voltage (V_{CE})

and the h-parameters are given by

- $h_{ix} = h_{ie}$, the input impedance of the transistor (corresponding to the base resistance r_{pi}).

- $h_{rx} = h_{re}$, represents the dependence of the transistor's I_B–V_{BE} curve on the value
  of V_{CE}. It is usually very small and is often neglected (assumed to be zero).

- $h_{fx} = h_{fe}$, the current-gain of the transistor. This parameter is often specified as
  h_{FE} or the DC current-gain (β_{DC}) in datasheets.

- $h_{ox} = 1/h_{oe}$, the output impedance of transistor. The parameter h_{oe} usually corresponds to the output admittance of the bipolar transistor and has to be inverted
  to convert it to an impedance.

As shown, the h-parameters have lower-case subscripts and hence signify AC conditions or analyses. For DC conditions they are specified in upper-case. For the CE topology, an approximate h-parameter model is commonly used which further simplifies the circuit analysis. For this the h_{oe} and h_{re} parameters are neglected (that is, they are set to infinity and zero, respectively). The h-parameter model as shown is suited to low-frequency, small-signal analysis. For high-frequency analyses the inter-electrode capacitances that are important at high frequencies must be added.

Etymology of h_{FE}

The *h* refers to its being an h-parameter, a set of parameters named for their origin in a *hybrid equivalent circuit* model. *F* is from *forward current amplification* also called the current gain. *E* refers to the transistor operating in a *common emitter* (CE) configuration. Capital letters used in the subscript indicate that h_{FE} refers to a direct current circuit.

Industry Models

The Gummel–Poon SPICE model is often used, but it suffers from several limitations. These have been addressed in various more advanced models: Mextram, VBIC, HICUM, Modella.

Applications

The BJT remains a device that excels in some applications, such as discrete circuit design, due to the very wide selection of BJT types available, and because of its high transconductance and output resistance compared to MOSFETs.

The BJT is also the choice for demanding analog circuits, especially for very-high-frequency applications, such as radio-frequency circuits for wireless systems.

High Speed Digital Logic

Emitter-coupled logic (ECL) use BJTs.

Bipolar transistors can be combined with MOSFETs in an integrated circuit by using a BiCMOS process of wafer fabrication to create circuits that take advantage of the application strengths of both types of transistor.

Amplifiers

The transistor parameters α and β characterizes the current gain of the BJT. It is this gain that allows BJTs to be used as the building blocks of electronic amplifiers. The three main BJT amplifier topologies are:

- Common emitter

- Common base

- Common collector

Temperature Sensors

Because of the known temperature and current dependence of the forward-biased base–emitter junction voltage, the BJT can be used to measure temperature by subtracting two voltages at two different bias currents in a known ratio.

Logarithmic Converters

Because base–emitter voltage varies as the logarithm of the base–emitter and collector–emitter currents, a BJT can also be used to compute logarithms and anti-logarithms. A diode can also perform these nonlinear functions but the transistor provides more circuit flexibility.

Vulnerabilities

Exposure of the transistor to ionizing radiation causes radiation damage. Radiation causes a buildup of 'defects' in the base region that act as recombination centers. The resulting reduction in minority carrier lifetime causes gradual loss of gain of the transistor.

Transistors have "Maximum Ratings", including Power ratings (essentially limited by self-heating), maximum collector and base currents (both continuous/DC ratings and peak), and Breakdown voltage ratings, beyond which the device may fail or at least perform badly.

In addition to normal breakdown ratings of the device, power BJTs are subject to a failure mode called secondary breakdown, in which excessive current and normal imperfections in the silicon die cause portions of the silicon inside the device to become disproportionately hotter than the others. The electrical resistivity of doped silicon, like other semiconductors, has a negative temperature coefficient, meaning that it conducts more current at higher temperatures. Thus, the hottest part of the die conducts the most current, causing its conductivity to increase, which then causes it to become progressively hotter again, until the device fails internally. The thermal runaway process associated with secondary breakdown, once triggered, occurs almost instantly and may catastrophically damage the transistor package.

If the emitter-base junction is reverse biased into avalanche or Zener mode and charge flows for a short period of time, the current gain of the BJT will be permanently degraded.

Common Base Amplifier

In electronics, a common base (also known as grounded-base) amplifier is one of three basic single-stage bipolar junction transistor (BJT) amplifier topologies, typically used as a current buffer or voltage amplifier.

In this circuit the emitter terminal of the transistor serves as the input, the collector the output, and the base is connected to ground, or "common", hence its name. The analogous field-effect transistor circuit is the common gate amplifier.

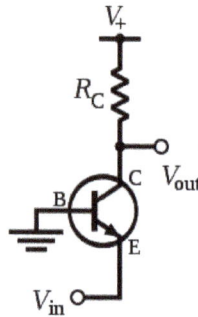

(a) Basic NPN common base circuit (neglecting biasing details).

Simplified Operation

As current is sunk from the emitter this provides potential difference so causing the transistor to conduct. The current conducted via the collector is proportional to the voltage across the base-emitter junction, accounting for the bias, as with other configurations.

Therefore, if no current is sunk at the emitter the transistor does not conduct.

Applications

This arrangement is not very common in low-frequency discrete circuits, where it is usually employed for amplifiers that require an unusually low input impedance, for example to act as a preamplifier for moving-coil microphones. However, it is popular in integrated circuits and in high-frequency amplifiers, for example for VHF and UHF, because its input capacitance does not suffer from the Miller effect, which degrades the bandwidth of the common emitter configuration, and because of the relatively high isolation between the input and output. This high isolation means that there is little feedback from the output back to the input, leading to high stability.

This configuration is also useful as a current buffer since it has a current gain of approximately unity. Often a common base is used in this manner, preceded by a common emitter stage. The combination of these two form the cascode configuration, which possesses several of the benefits of each configuration, such as high input impedance and isolation.

Low-frequency Characteristics

At low frequencies and under small-signal conditions, the circuit in Figure A can be represented by that in Figure B, where the hybrid-pi model for the BJT has been employed. The input signal is represented by a Thévenin voltage source, v_s, with a series resistance R_s and the load is a resistor R_L. This circuit can be used to derive the following characteristics of the common base amplifier.

	Definition	Expression	Approximate expression	Conditions
Open-circuit voltage gain	$A_v = \dfrac{v_o}{v_i}\Big|_{R_L=\infty}$	$\dfrac{(g_m r_o +1)R_C}{R_C + r_o}$	$g_m R_C$	$r_o \gg R_C$
Short-circuit current gain	$A_i = \dfrac{i_o}{i_i}\Big|_{R_L=0}$	$\dfrac{r_\pi + \beta r_o}{r_\pi + (\beta+1)r_o}$	1	$\beta \gg 1$
Input resistance	$R_{in} = \dfrac{v_i}{i_i}$	$\dfrac{(r_o + R_C \| R_L)r_E}{r_o + r_E + \dfrac{R_C \| R_L}{\beta+1}}$	$r_e \left(\approx \dfrac{1}{g_m}\right)$	$r_o \gg R_C \| R_L$ $(\beta \gg 1)$
Output resistance	$R_{out} = \dfrac{v_o}{-i_o}\Big|_{v_s=0}$	$R_C \| \left(\dfrac{[1+g_m}{(r_\pi \| R_S)]r_o + r_\pi \| R_S}\right)$	$R_C \| r_o$ $R_C \| \left(r_o\left[1+g_m\left(r_\pi \| R_S\right)\right]\right)$	$R_S \ll r_E$ $R_S \gg r_E$

Note: Parallel lines (||) indicate components in parallel.

In general the overall voltage/current gain may be substantially less than the open/short circuit gains listed above (depending on the source and load resistances) due to the loading effect.

Active Loads

For voltage amplification, the range of allowed output voltage swing in this amplifier is tied to voltage gain when a resistor load R_C is employed, as in Figure A. That is, large voltage gain requires large R_C, and that in turn implies a large DC voltage drop across R_C. For a given supply voltage, the larger this drop, the smaller the transistor V_{CB} and the less output swing is allowed before saturation of the transistor occurs, with resultant distortion of the output signal. To avoid this situation, an active load can be used, for example, a current mirror. If this choice is made, the value of R_C in the table above is replaced by the small-signal output resistance of the active load, which is generally at least as large as the r_o of the active transistor in Figure A. On the other hand, the DC voltage drop across the active load is a fixed low value (the compliance voltage of the active load), much less than the DC voltage drop incurred for comparable gain using a resistor R_C. That is, an active load imposes less restriction on the output voltage swing. Notice that active load or not, large AC gain still is coupled to large AC output resistance, which leads to poor voltage division at the output except for large loads $R_L \gg R_{out}$.

For use as a current buffer, gain is not affected by R_C, but output resistance is. Because of the current division at the output, it is desirable to have an output resistance for the buffer much larger than the load R_L being driven so large signal currents can be delivered to a load. If a resistor R_C is used, as in Figure A, a large output resistance is coupled to a large R_C, again limiting the signal swing at the output. (Even though current is delivered to the load, usually a large current signal into the load implies a large voltage swing across the load as well.) An active load provides high AC output resistance with much less serious impact upon the amplitude of output signal swing.

Overview of Characteristics

Several example applications are described in detail below. A brief overview follows.

- The amplifier input impedance R_{in} looking into the emitter node is very low, given approximately by

$$R_{in} = r_E = \frac{V_T}{I_E},$$

where V_T is the thermal voltage and I_E is the DC emitter current.

For example, for V_T = 26 mV and I_E = 10 mA, rather typical values, R_{in} = 2.6 Ω. If I_E is reduced to increase R_{in}, there are other consequences like lower transconductance, higher output resistance and lower β that also must be considered. A practical solution to this low-input-impedance problem is to place a common emitter stage at the input to form a cascode amplifier.

- Because the input impedance is so low, most signal sources have larger source impedance than the common base amplifier R_{in}. The consequence is that the source delivers a *current* to the input rather than a voltage, even if it is a voltage source. (According to Norton's theorem, this current is approximately $i_{in} = v_S / R_S$). If the output signal also is a current, the amplifier is a current buffer and delivers the same current as is input. If the output is taken as a voltage, the amplifier is a transresistance amplifier, and delivers a voltage dependent on the load impedance, for example $v_{out} = i_{in} R_L$ for a resistor load R_L much smaller in value than the amplifier output resistance R_{out}. That is, the voltage gain in this case (explained in more detail below) is:

$$V_{out} = i_{in} R_L = V_S \frac{R_L}{R_S} \qquad \rightarrow A_v = \frac{V_{out}}{V_S} = \frac{R_L}{R_S}.$$

Note for source impedances such that $R_S \gg r_E$ the output impedance approaches R_{out} = $R_C \,||\, [\, g_m (r_\pi \,||\, R_S) r_o \,]$.

- For the special case of very low impedance sources, the common base amplifier does work as a voltage amplifier, one of the examples discussed below. In this case (explained in more detail below), when $R_S \ll r_E$ and $R_L \ll R_{out}$, the voltage gain becomes:

$$A_v = \frac{V_{out}}{V_S} = \frac{R_L}{r_E} \approx g_m R_L$$

where $g_m = I_C / V_T$ is the transconductance. Notice that for low source impedance, $R_{out} = r_o \,||\, R_C$.

- The inclusion of r_O in the hybrid-pi model predicts reverse transmission from the amplifiers output to its input, that is the amplifier is bilateral. One consequence of this is that the input/output impedance is affected by the load/source termination impedance, hence, for example, the output resistance, R_{out}, may vary over the range $r_o \,||\, R_C \le R_{out} \le (\beta + 1) \, r_o \,||\, R_C$ depending on the source resistance, R_S. The amplifier can be approximated as unilateral when neglect of r_O is accurate (valid for low gains and low to moderate load resistances), simplifying the analysis. This approximation often is made in discrete designs, but may be less accurate in RF circuits, and in integrated circuit designs where active loads normally are used.

Voltage Amplifier

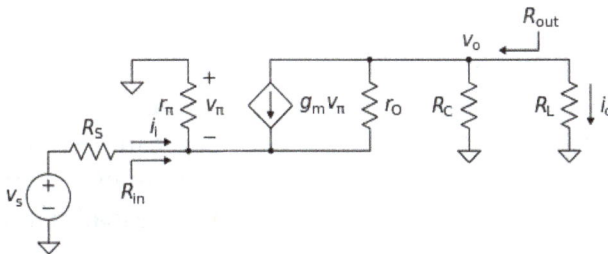

(b) Small-signal model for calculating various parameters; Thévenin voltage source as signal.

For the case when the common base circuit is used as a voltage amplifier, the circuit is shown in the above figure.

The output resistance is large, at least $R_C \,||\, r_O$, the value which arises with low source impedance ($R_S \ll r_E$). A large output resistance is undesirable in a voltage amplifier, as it leads to poor voltage division at the output. Nonetheless, the voltage gain is appreciable even for small loads: according to the table, with $R_S = r_E$ the gain is $A_v = g_m R_L / 2$. For larger source impedances, the gain is determined by the resistor ratio R_L / R_S, and not by the transistor properties, which can be an advantage where insensitivity to temperature or transistor variations is important.

An alternative to the use of the hybrid-pi model for these calculations is a general technique based upon two-port networks. For example, in an application like this one where voltage is the output, a g-equivalent two-port could be selected for simplicity, as it uses a voltage amplifier in the output port.

For R_S values in the vicinity of r_E the amplifier is transitional between voltage amplifier and current buffer. For $R_S \gg r_E$ the driver representation as a Thévenin source should be replaced by representation with a Norton source. The common base circuit stops behaving like a voltage amplifier and behaves like a current follower, as discussed next.

Current Follower

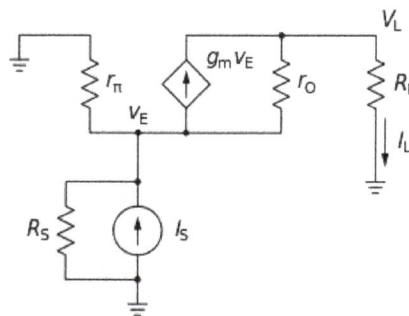

(c) Common base circuit wi th Norton driver; R_C is omitted because an active load is assumed with infinite small-signal output resistance

Figure above shows the common base amplifier used as a current follower. The circuit signal is provided by an AC Norton source (current I_S, Norton resistance R_S) at the input, and the circuit has a resistor load R_L at the output.

As mentioned earlier, this amplifier is bilateral as a consequence of the output resistance r_o, which connects the output to the input. In this case the output resistance is large even in the worst case (it is at least $r_o \parallel R_C$ and can become $(\beta + 1) r_o \parallel R_C$ for large R_S). Large output resistance is a desirable attribute of a current source because favorable current division sends most of the current to the load. The current gain is very nearly unity as long as $R_S \gg r_E$.

An alternative analysis technique is based upon two-port networks. For example, in an application like this one where current is the output, an h-equivalent two-port is selected because it uses a current amplifier in the output port.

The common base amplifier circuit is shown in the figure below. The V_{EE} source forward biases the emitter diode and V_{CC} source reverse biased collector diode. The ac source vin is connected to emitter through a coupling capacitor so that it blocks dc. This ac voltage produces small fluctuation in currents and voltages. The load resistance R_L is also connected to collector through coupling capacitor so the fluctuation in collector base voltage will be observed across R_L.

The dc equivalent circuit is obtained by reducing all ac sources to zero and opening all capacitors. The dc collector current is same as I_E and V_{CB} is given by

$$V_{CB} = V_{CC} - I_C R_C$$

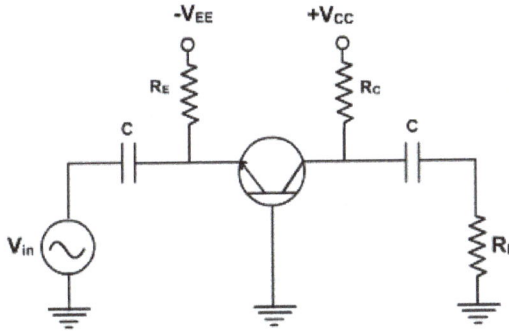

These current and voltage fix the Q point. The ac equivalent circuit is obtained by reducing all dc sources to zero and shorting all coupling capacitors. r'_e represents the ac resistance of the diode as shown in the above figure.

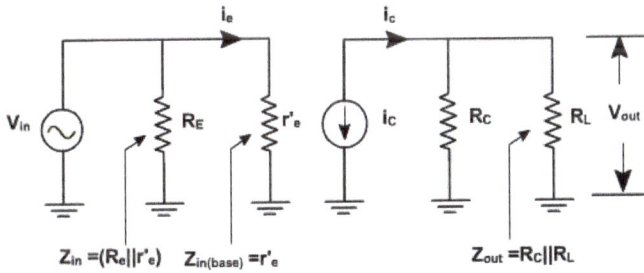

$$Z_{in} = (R_e || r'_e) \quad Z_{in(base)} = r'_e \qquad\qquad Z_{out} = R_C || R_L$$

Figure below, shows the diode curve relating I_E and V_{BE}. In the absence of ac signal, the transistor operates at Q point (point of intersection of load line and input characteristic). When the ac signal is applied, the emitter current and voltage also change. If the signal is small, the operating point swings sinusoidally about Q point (A to B).

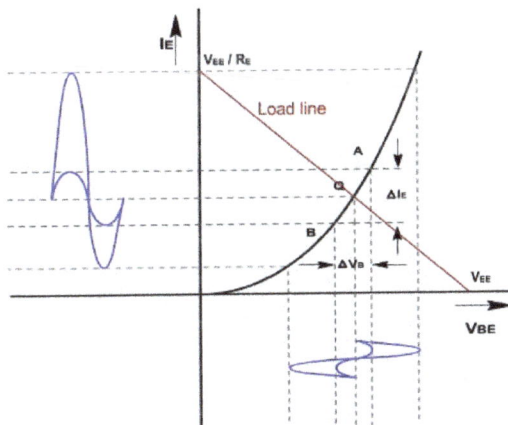

If the ac signal is small, the points A and B are close to Q, and arc A B can be approximated by a straight line and diode appears to be a resistance given by

$$r_e' = \frac{\Delta V_{BE}}{\Delta |E|}\bigg|_{\text{small change}}$$

$$= \frac{V_{be}}{i_e} = \frac{\text{ac voltage across base and emitter}}{\text{a current through emitter}}$$

If the input signal is small, input voltage and current will be sinusoidal but if the input voltage is large then current will no longer be sinusoidal because of the non linearity of diode curve. The emitter current is elongated on the positive half cycle and compressed on negative half cycle. Therefore the output will also be distorted.

r_e' is the ratio of ΔV_{BE} and ΔI_E and its value depends upon the location of Q. Higher up the Q point small will be the value of r_e' because the same change in V_{BE} produces large change in I_E. The slope of the curve at Q determines the value of r_e'. From calculation it can be proved that.

$$r_e' = 25mV/I_E$$

Proof:

In general, the current through a diode is given by

$$I = I_{CO}(e^{\frac{qV}{KT}} - 1)$$

Where q is he charge on electron, V is the drop across diode, T is the temperature and K is a constant.

On differentiating w.r.t V, we get,

$$\frac{dI}{dV} = I_{co} * e^{\frac{qV}{KT}} * \frac{q}{KT}$$

The value of (q / KT) at 25°C is approximately 40.

Therefore, $$\frac{dI}{dV} = 40 * I_{CO} * e^{\frac{qV}{KT}}$$

$$= 40 * (I + I_{CO})$$

or, $$\frac{dV}{dI} = \frac{1}{40 * (I + I_{CO})} \approx \frac{1}{40 * I}$$

Therefore, ac resistance of the emitter diode $= \dfrac{dV}{dI} = \dfrac{25mV}{I}$ Ohms

To a close approximation the small changes in collector current equal the small changes in emitter current. In the ac equivalent circuit, the current 'i_c' is shown upward because if 'i_e' increases, then 'i_c' also increases in the same direction.

Voltage gain:

Since the ac input voltage source is connected across r_e'. Therefore, the ac emitter current is given by

$$i_e = V_{in}/r_e'$$

or, $V_{in} = i_e r_e'$

The output voltage is given by $V_{out} = i_C (R_C \parallel R_L)$

Therefore, voltage gain $A_V = \dfrac{V_{out}}{V_{in}} = \dfrac{(R_C \parallel R_L)}{r_e'}$

$$= \dfrac{rc}{r}$$

Under open circuit condition $V_{out} = i_C R_C$

Therefore, voltage gain in open circuit condition $= A_V = \dfrac{R_C}{r_e'}$

Example-1

Find the voltage gain and output of the amplifier shown in the below figure, if input voltage is 1.5mV.

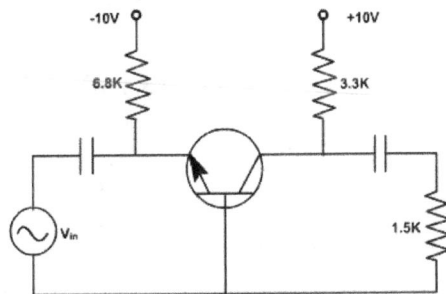

Solution:

The emitter dc current I E is given by $I_E = \dfrac{10 - 0.7}{6.8k} = 1.37mA$

Therefore, emitter ac resistance $= A_V = \dfrac{r_c}{r_e'} = \dfrac{3.3k \,||\, 1.5k}{18.2\Omega}$

or, $\qquad A_v = 56.6$

and, $V_{out} = 1.5 \times 56.6 = 84.9 mV$

Example-2

Repeat example-1 if ac source has resistance R s = 100 W .

Solution:

The ac equivalent circuit with ac source resistance is shown in the below figure.

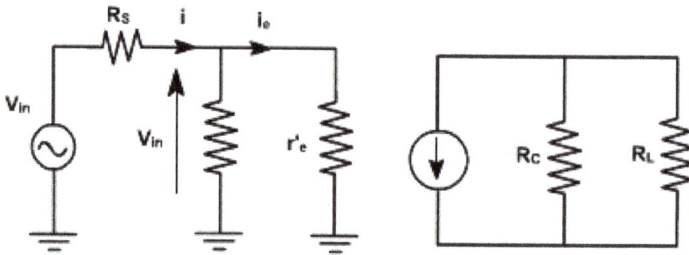

The emitter ac current is given by $i_e = \dfrac{V_{in}}{R_s + (R_E \,||\, r_e')} \times \dfrac{R_E}{R_E + r_e'}$

or, $\qquad i_e = \dfrac{V_{in}}{(R_S + r_e')R_E R_S r_e')} \times R_E ; \dfrac{V_{in}}{R_S + r_e'}$

Therefore, voltage gain of the amplifier $= A_V = \dfrac{V_{out}}{V_{in}} = \dfrac{i_c r_c}{i_e(R_S + r_e')} = \dfrac{r_c}{R_S + r_e'}$

$$A_V = \dfrac{3.3k \,||\, 1.5k}{100\Omega + 18.2\Omega} = 8.71$$

and, $\qquad\qquad V_{out} = 1.5 \times 8.71 = 13.1 mV$

Common Emitter Amplifier

In electronics, a common emitter amplifier is one of three basic single-stage bipolar-junction-transistor (BJT) amplifier topologies, typically used as a voltage amplifier.

In this circuit the base terminal of the transistor serves as the input, the collector is the output, and the emitter is *common* to both (for example, it may be tied to ground reference or a power supply rail), hence its name. The analogous FET circuit is the common source amplifier, and the analogous tube circuit is the common cathode amplifier.

Emitter Degeneration

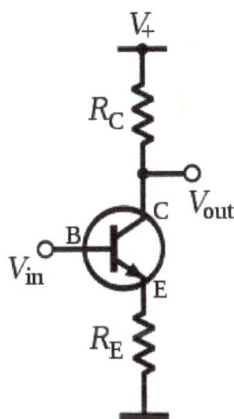

Adding an emitter resistor decreases gain, but increases linearity and stability

Common emitter amplifiers give the amplifier an inverted output and can have a very high gain that may vary widely from one transistor to the next. The gain is a strong function of both temperature and bias current, and so the actual gain is somewhat unpredictable. Stability is another problem associated with such high gain circuits due to any unintentional positive feedback that may be present.

Other problems associated with the circuit are the low input dynamic range imposed by the small-signal limit; there is high distortion if this limit is exceeded and the transistor ceases to behave like its small-signal model. One common way of alleviating these issues is with *emitter degeneration*. This refers to the addition of a small resistor (or any impedance) between the emitter and the common signal source (e.g., the ground reference or a power supply rail). This impedance R_E reduces the overall transconductance $G_m = g_m$ of the circuit by a factor of $g_m R_E + 1.677$, which makes the voltage gain.

$$A_v \triangleq \frac{V_{out}}{v_{in}} = \frac{-g_m R_C}{g_m R_E + 1} \approx -\frac{R_C}{R_E} \qquad (\text{where} \quad g_m R_E \gg 1).$$

The voltage gain depends almost exclusively on the ratio of the resistors R_C / R_E rather than the transistor's intrinsic and unpredictable characteristics. The distortion and stability characteristics of the circuit are thus improved at the expense of a reduction in gain.

(While this is often described as "negative feedback", as it reduces gain, raises input impedance, and reduces distortion, it predates the invention of negative feedback and does not reduce output impedance or increase bandwidth, as true negative feedback would do.)

Characteristics

At low frequencies and using a simplified hybrid-pi model, the following small-signal characteristics can be derived.

	Definition	Expression (with emitter degeneration)	Expression (without emitter degeneration, i.e., $R_E = 0$)
Current gain	$A_i \triangleq \dfrac{i_{out}}{i_{in}}$	β	β
Voltage gain	$A_v \triangleq \dfrac{v_{out}}{v_{in}}$	$-\dfrac{\beta R_C}{r_\pi + (\beta + 1)R_E}$	$-g_m R_C$
Input impedance	$r_{in} \triangleq \dfrac{v_{in}}{i_{in}}$	$r_\pi + (\beta + 1)R_E$	r_π
Output impedance	$r_{out} \triangleq \dfrac{v_{out}}{i_{out}}$	R_C	R_C

If the emitter degeneration resistor is not present, then $R_E = 0\Omega$, and the expressions effectively simplify to the ones given by the rightmost column (note that the voltage gain is an ideal value; the actual gain is somewhat unpredictable). As expected, when R_E is increased, the input impedance is increased and the voltage gain A_v is reduced.

Bandwidth

The bandwidth of the common-emitter amplifier tends to be low due to high capacitance resulting from the Miller effect. The parasitic base-collector capacitance C_{CB} appears like a larger parasitic capacitor $C_{CB}(1 - A_v)$ (where A_v is negative) from the base to ground. This large capacitor greatly decreases the bandwidth of the amplifier as it makes the time constant of the parasitic input RC filter $r_s(1 - A_v)C_{CB}$ where r_s is the output impedance of the signal source connected to the ideal base.

The problem can be mitigated in several ways, including:

- Reduction of the voltage gain magnitude $|A_v|$ (e.g., by using emitter degeneration).

- Reduction of the output impedance r_s of the signal source connected to the base (e.g., by using an emitter follower or some other voltage follower).

- Using a cascode configuration, which inserts a low input impedance current buffer (e.g. a common base amplifier) between the transistor's collector and the load. This configuration holds the transistor's collector voltage roughly constant, thus making the base to collector gain zero and hence (ideally) removing the Miller effect.

- Using a differential amplifier topology like an emitter follower driving a grounded-base amplifier; as long as the emitter follower is truly a common-collector amplifier, the Miller effect is removed.

The Miller effect negatively affects the performance of the common source amplifier in the same way (and has similar solutions).When an AC signal is applied to the transistor amplifier it causes the base voltage VB to fluctuate in value at the AC signal. The positive half of the applied signal will cause an increase in the value of VB this turn will increase the base current IB and cause a corresponding increase in emitter current IE and collector current IC. As a result, the collector emitter voltage will be reduced because of the increase voltage drop across RL. The negative alternation of an AC signal will cause a decrease in IB this action then causes a corresponding decrease in IE through RL. The output signal of a common- emitter amplifier is therefore 180 degrees out of phase with the input signal.

It is also named common- emitter amplifier because the emitter of the transistor is common to both the input circuit and output circuit. The input signal is applied across the ground and the base circuit of the transistor. The output signal appears across ground and the collector of the transistor. Since the emitter is connected to the ground, it is common to signals, input and output.

The common- emitter circuit is the most widely used of junction, transistor amplifiers. As compared with the common- base connection, it has higher input impedance and lower output impedance. A single power supply is easily used for biasing. In addition, higher voltage and power gains are usually obtained for common- emitter (CE) operation.

Current gain in the common emitter circuit is obtained from the base and the collector circuit currents. Because a very small change in base current produces a large change in collector current, the current gain (β) is always greater than unity for the common-emitter circuit, a typical value is about 50.

Applications

Low Frequency Voltage Amplifier

A typical example of the use of a common-emitter amplifier is shown in the figure below.

The input capacitor C removes any constant component of the input, and the resistors R_1 and R_2 bias the transistor so that it will remain in active mode for the entire range of the input. The output is an inverted copy of the AC-component of the input that has been amplified by the ratio R_C/R_E and shifted by an amount determined by all four resistors. Because R_C is often large, the output impedance of this circuit can be prohibitively high. To alleviate this problem, R_C is kept as low as possible and the amplifier is followed by a voltage buffer like an emitter follower.

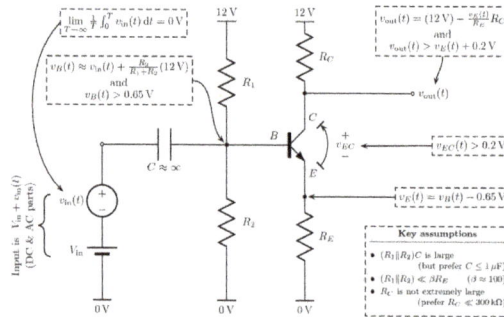

Single-ended *npn* common-emitter amplifier with emitter degeneration. The AC-coupled circuit acts as a level-shifter amplifier. Here, the base–emitter voltage drop is assumed to be 0.65 Volts.

Radio

Common-emitter amplifiers are also used in radio frequency circuits, for example to amplify faint signals received by an antenna. In this case it is common to replace the load resistor with a tuned circuit. This may be done to limit the bandwidth to a narrow band centered around the intended operating frequency. More importantly it also allows the circuit to operate at higher frequencies as the tuned circuit can be used to resonate any inter-electrode and stray capacitances, which normally limit the frequency response. Common emitters are also commonly used as low-noise amplifiers.

The common emitter configuration of BJT is shown in the figure below.

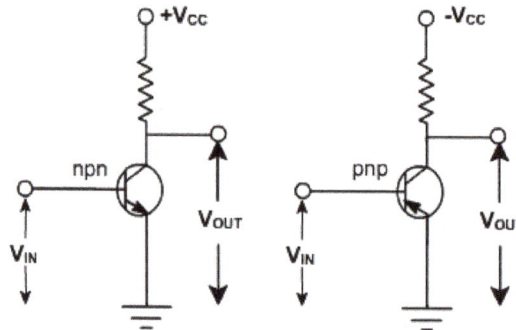

In C.E. configuration the emitter is made common to the input and output. It is also referred to as grounded emitter configuration. It is most commonly used configuration. In this, base current and output voltages are taken as impendent parameters and input voltage and output current as dependent parameters

$$V_{BE} = f_1(I_B, V_{CE})$$

$$I_C = f_2(I_B, V_{CE})$$

Input Characteristic:

The curve between I_B and V_{BE} for different values of V_{CE} are shown in the figure below. Since the base emitter junction of a transistor is a diode, therefore the characteristic is similar to diode one. With higher values of V_{CE} collector gathers slightly more electrons and therefore base current reduces. Normally this effect is neglected. (Early effect). When collector is shorted with emitter then the input characteristic is the characteristic of a forward biased diode when V_{BE} is zero and I_B is also zero.

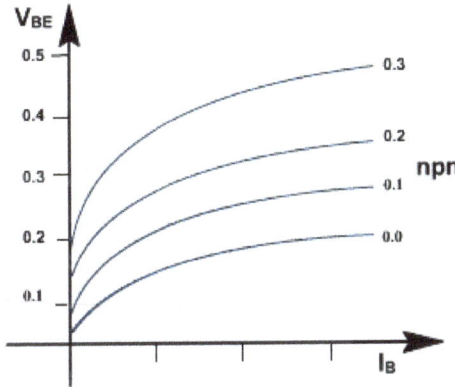

Output Characteristic:

The output characteristic is the curve between V_{CE} and I_C for various values of I_B. For fixed value of I_B and is shown in the figure below. For fixed value of I_B, I_C is not varying much dependent on V_{CE} but slopes are greater than CE characteristic. The output characteristics can again be divided into three parts.

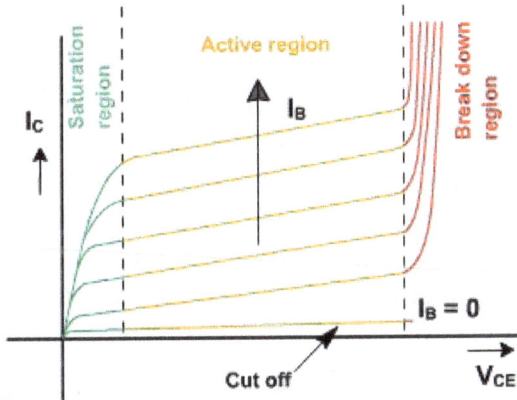

(1) Active Region

In this region collector junction is reverse biased and emitter junction is forward biased. It is the area to the right of V_{CE} = 0.5 V and above I_B= 0. In this region transistor current responds most sensitively to I_B. If transistor is to be used as an amplifier, it must operate in this region.

$$I_E = I_C + I_B$$

Since, $I_C = I_{CO} + \alpha_{dc} I_E$

$$I_C = I_{CO} + \alpha_{dc}(I_C + I_B)$$

or $(1 - \alpha_{dc})I_C = \alpha_{dc}I_B + I_{CO}$

or $I_C = \left(\dfrac{\alpha_{dc}}{1 - \alpha_{dc}}\right)I_B + \left(\dfrac{1}{1 - \alpha_{dc}}\right)I_{CO}$

Let, $\beta_{dc} = \dfrac{\alpha_{dc}}{1 - \alpha_{dc}}$

$$\therefore I_C = (1 + \beta_{dc})I_{CO} + \beta_{dc}I_B$$

β_{dc} is defined as current gain of the transistor is given by

$$\beta_{dc} = \dfrac{I_C - I_{CO}}{I_B + I_{CO}}$$

If α_{dc} is truly constant then I_C would be independent of V_{CE}. But because of early effect, α_{dc} increases by 0.1% (0.001) e.g. from 0.995 to 0.996 as V_{CE} increases from a few volts to 10V. Then β_{dc} increases from 0.995 / (1-0.995) = 200 to 0.996 / (1-0.996) = 250 or about 25%. This shows that small change in α reflects large change in β. Therefore the curves are subjected to large variations for the same type of transistors.

(2) Cut Off

Cut off in a transistor is given by $I_B = 0$, $I_C = I_{CO}$. A transistor is not at cut off if the base current is simply reduced to zero (open circuited) under this condition,

$$I_C = I_E = I_{CO} / (1 - \alpha_{dc}) = I_{CEO}$$

The actual collector current with base open is designated as I_{CEO}. Since even in the neighborhood of cut off, α_{dc} may be as large as 0.9 for Ge, then $I_C = 10\, I_{CO}$(approximately), at zero base current. Accordingly in order to cut off transistor it is not enough to reduce I_B to zero, but it is necessary to reverse bias the emitter junction slightly. It is found that reverse voltage of 0.1 V is sufficient for cut off a transistor. In Si, the α_{dc} is very nearly equal to zero, therefore, $I_C = I_{CO}$. Hence even with $I_B = 0$, $I_C = I_E = I_{CO}$ so that transistor is very close to cut off.

Cut off means $I_E = 0$, $I_C = I_{CO}$, $I_B = -I_C = -I_{CO}$, and V_{BE} is a reverse voltage whose magnitude is of the order of 0.1 V for Ge and 0 V for Si.

Reverse Collector Saturation Current ICBO

When in a physical transistor emitter current is reduced to zero, then the collector current is known as I_{CBO} (approximately equal to I_{CO}). Reverse collector saturation current

I_{CBO} also varies with temperature, avalanche multiplication and variability from sample to sample. Consider the circuit shown in the figure below. V_{BB} is the reverse voltage applied to reduce the emitter current to zero.

$$I_E = 0, \qquad I_B = -I_{CBO}$$

If we require, $V_{BE} = -0.1\,V$

Then $-V_{BB} + I_{CBO}\,R_B < -0.1\,V$

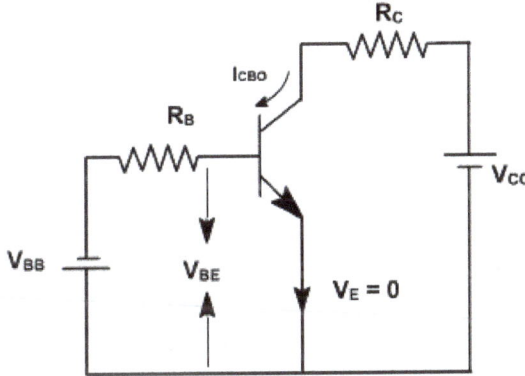

If $R_B = 100\,K$, $I_{CBO} = 100\,m A$, Then V_{BB} must be 10.1 Volts. Hence transistor must be capable to withstand this reverse voltage before breakdown voltage exceeds.

(3) Saturation Region

In this region both the diodes are forward biased by at least cut in voltage. Since the voltage V_{BE} and V_{BC} across a forward is approximately 0.7 V therefore, $V_{CE} = V_{CB} + V_{BE} = -V_{BC} + V_{BE}$ is also few tenths of volts. Hence saturation region is very close to zero voltage axis, where all the current rapidly reduces to zero. In this region the transistor collector current is approximately given by V_{CC} / R_C and independent of base current. Normal transistor action is last and it acts like a small ohmic resistance.

Large Signal Current Gain β_{dc} :-

The ratio I_C/I_B is defined as transfer ratio or large signal current gain β_{dc}

$$\beta_{dc} = \frac{I_C}{I_B}$$

Where I_C is the collector current and I_B is the base current. The β_{dc} is an indication if how well the transistor works. The typical value of β_{dc} varies from 50 to 300.

In terms of h parameters, β_{dc} is known as dc current gain and in designated h_{fE} ($\beta_{dc} = h_{fE}$). Knowing the maximum collector current and β_{dc} the minimum base current can be found which will be needed to saturate the transistor.

$$I_{C(max)} = \frac{V_{CC} - V_{CE(sat)}}{R_C} = I_{C(sat)}$$

$$I_{B(min)} = \frac{I_{C(sat)}}{\beta_{dc}}$$

This expression of β_{dc} is defined neglecting reverse leakage current (I_{CO}).

Taking reverse leakage current (I_{CO}) into account, the expression for the β_{dc} can be obtained as follows:

β_{dc} in terms of α_{dc} is given by

$$\beta_{dc} = \frac{\alpha_{dc}}{1 - \alpha_{dc}}$$

$$= \frac{\dfrac{I_C - I_{CO}}{I_E}}{1 - \dfrac{I_C - I_{CO}}{I_E}} = \frac{I_C - I_{CO}}{I_E - I_C + I_{CO}}$$

$$= \frac{I_C - I_{CO}}{I_B + I_{CO}}$$

Since, $I_{CO} = I_{CBO}$

$$\therefore \beta_{dc} = \frac{I_C - I_{CBO}}{I_B + I_{CBO}}$$

Cut off of a transistor means $I_E = 0$, then $I_C = I_{CBO}$ and $I_B = -I_{CBO}$. Therefore, the above expression β_{dc} gives the collector current increment to the base current change form cut off to I_B and hence it represents the large signal current gain of all common emitter transistor.

Biasing

Biasing in electronics means establishing predetermined voltages or currents at various points of an electronic circuit for the purpose of establishing proper operating conditions in electronic components. Many electronic devices such as transistors and vacuum tubes, whose function is processing time-varying (AC) signals also require a steady (DC) current or voltage to operate correctly — a *bias*. The AC signal applied to them is superposed on this DC bias current or voltage. The operating point of a device, also known as bias point, quiescent point, or Q-point, is the steady-state (DC) voltage or current at a specified terminal of an active device (a transistor or vacuum tube) with no input signal applied. A *bias circuit* is a portion of the device's circuit which supplies this steady current or voltage.

The term is also used for an alternating current (AC) signal applied to some electronic devices which is similarly required for correct operation, such as the tape bias signal applied to magnetic recording heads used in magnetic tape recorders.

Overview

In electronic engineering, the term *bias* has the following meanings:

1. A systematic deviation of a value from a reference value.

2. The amount by which the average of a set of values departs from a reference value.

3. Electrical, mechanical, magnetic, or other force (field) applied to a device to establish a reference level to operate the device.

4. In telegraph signaling systems, the development of a positive or negative DC voltage at a point on a line that should remain at a specified reference level, such as zero.

Note: A bias may be applied or produced by (i) the electrical characteristics of the line, (ii) the terminal equipment, and (iii) the signaling scheme.

In electronics, *bias* usually refers to a fixed DC voltage or current applied to a terminal of an electronic component such as a diode, transistor or vacuum tube in an circuit in which alternating current (AC) signals are also present, in order to establish proper operating conditions for the device. For example, a bias voltage is applied to a transistor in an electronic amplifier to allow the transistor to operate in a particular region of its transconductance curve. For vacuum tubes, a grid bias voltage is often applied to the grid electrodes for the same reason.

In magnetic tape recording, the term *bias* is also used for a high-frequency signal added to the audio signal applied to the recording head, to improve the quality of the recording on the tape. This is called tape bias.

Bias is used in direct broadcast satellites such as DirecTV and Dish Network, the integrated receiver/decoder (IRD) box actually powers the feedhorn or low-noise block converter (LNB) receiver mounted on the dish arm. This bias is changed from a lower voltage to a higher voltage to select the polarization of the LNB, so that it receives signals that are polarized either horizontally or vertically, thereby allowing it to receive twice as many channels.

We still need to determine the optimal values for the DC biasing in order to choose resistors, etc. This bias point is called the quiescent or Q-point as it gives the values of the voltages when no input signal is applied. To determine the Q-point we need to look at the range of values for which the transistor is in the active region.

Importance in Linear Circuits

Linear circuits involving transistors typically require specific DC voltages and currents for correct operation, which can be achieved using a biasing circuit. As an example of the need for careful biasing, consider a transistor amplifier. In linear amplifiers, a small input signal gives larger output signal without any change in shape (low distortion): the input signal causes the output signal to vary up and down about the Q-point in a manner strictly proportional to the input. However, because a transistor is nonlinear, the transistor amplifier only approximates linear operation. For low distortion, the transistor must be biased so the output signal swing does not drive the transistor into a region of extremely nonlinear operation. For a bipolar transistor amplifier, this requirement means that the transistor must stay in the active mode, and avoid cut-off or saturation. The same requirement applies to a MOSFET amplifier, although the terminology differs a little: the MOSFET must stay in the active mode (or saturation mode), and avoid cut-off or ohmic operation (or triode mode).

Bipolar Junction Transistors

For bipolar junction transistors the bias point is chosen to keep the transistor operating in the active mode, using a variety of circuit techniques, establishing the Q-point DC voltage and current. A small signal is then applied on top of the Q-point bias voltage, thereby either modulating or switching the current, depending on the purpose of the circuit.

The quiescent point of operation is typically near the middle of the DC load line. The process of obtaining a certain DC collector current at a certain DC collector voltage by setting up the operating point is called biasing.

After establishing the operating point, when an input signal is applied, the output signal should not move the transistor either to saturation or to cut-off. However, this unwanted shift still might occur, due to the following reasons:

1. Parameters of transistors depend on junction temperature. As junction temperature increases, leakage current due to minority charge carriers (I_{CBO})(collector base current with emitter open) increases. As I_{CBO} increases, I_{CEO}(collector emitter current with base open) also increases, causing an increase in collector current I_C. This produces heat at the collector junction. This process repeats, and, finally, the Q-point may shift into the saturation region. Sometimes, the excess heat produced at the junction may even burn the transistor. This is known as thermal runaway.

2. When a transistor is replaced by another of the same type, the Q-point may shift, due to changes in parameters of the transistor, such as *current gain* (β) which varies slightly for each unique transistor.

To avoid a shift of Q-point, bias-stabilization is necessary. Various biasing circuits can be used for this purpose.

Vacuum Tubes

Grid bias is a DC voltage applied to electron tubes (or valves in British English) with three electrodes or more, such as triodes. The control grid (usually the first grid) of these devices is used to control the electron flow from the heated cathode to the positively charged anode. Bias point in small-signal applications is set to minimize distortion and achieve sufficiently low power draw. In high-power applications, biasing is typically set for maximum available output power or voltage, with a secondary target of *either* low distortion *or* high efficiency.

- In a typical voltage amplifier, including power stages of most audio power amplifiers, DC bias voltage is negative relative to cathode potential. Instant grid voltage (sum of DC bias and AC input signal) should never rise above cathode potential to prevent grid-to-cathode currents that overload preceding amplifier stages and may cause severe even-order distortion. High transconductance tubes develop significant grid currents even with small negative bias; in these cases, maximum instant voltage ceiling is lowered to -1.0..-0.5 Volt.

- High efficiency Class B+ push-pull amplifiers operate at higher bias points (near zero or even positive values). These designs take care of grid currents through the use of cathode followers or interstage transformers easing current load on the driver stages, and deep negative feedback to minimize distortion.

- High power transmitter tubes (oscillators and modulators) are frequently positively biased to maximize radio frequency output. Distortion is minimized by using band-pass filter loads tuned to the desired radio frequency.

Bias voltage is obtained through:

- An external voltage source (fixed bias) - a battery or a dedicated DC power supply. When the cathode potential is raised above ground (as in cascode circuits), bias voltage is obtained by tapping into main (positive) plate power supply.

- Automatic bias or *self bias* - using a cathode resistor to raise cathode potential above grid (tied to ground) and stabilize plate current;

- Grid leak bias - diverting DC grid current through a high value grid resistor, as used in the grid-leak detector.

Microphones

Electret microphone elements typically include a junction field-effect transistor as an impedance converter to drive other electronics within a few meters of the microphone. The operating current of this JFET is typically 0.1 to 0.5 mA and is often referred to as bias, which is different from the phantom power interface which supplies 48 volts to operate the backplate of a traditional condenser microphone. Electret microphone bias is sometimes supplied on a separate conductor.

Biasing Circuit Techniques

Fixed Bias or Base Bias

In order for a transistor to amplify, it has to be properly biased. This means forward biasing the base emitter junction and reverse biasing collector base junction. For linear amplification, the transistor should operate in active region (If I_E increases, I_C increases, V_{CE} decreases proportionally).

The source V_{BB}, through a current limit resistor R_B forward biases the emitter diode and V_{CC} through resistor R_C (load resistance) reverse biases the collector junction as shown in the figure below.

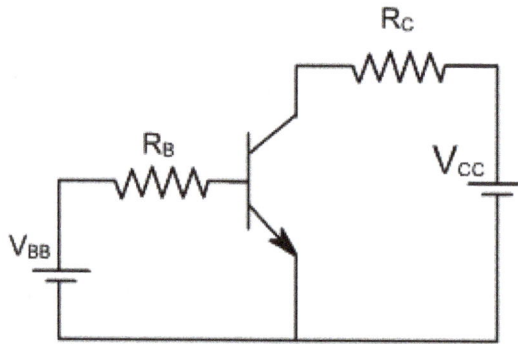

The dc base current through R_B is given by

$$I_B = (V_{BB} - V_{BE}) / R_B$$

or $$V_{BE} = V_{BB} - I_B R_B$$

Normally V_{BE} is taken 0.7V or 0.3V. If exact voltage is required, then the input characteristic (I_B vs V_{BE}) of the transistor should be used to solve the above equation. The load line for the input circuit is drawn on input characteristic. The two points of the load line can be obtained as given below

For $I_B = 0$, $V_{BE} = V_{BB}$.

and   For $V_{BE} = 0$, $I_B = V_{BB}/ R_B$.

The intersection of this line with input characteristic gives the operating point Q as shown in the figure below. If an ac signal is connected to the base of the transistor, then variation in V_{BE} is about Q - point. This gives variation in I_B and hence I_C.

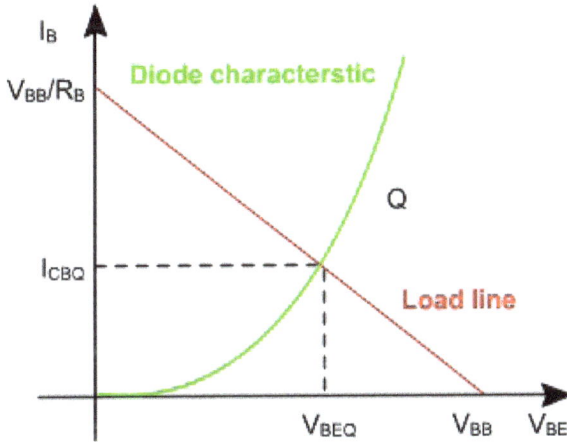

In the output circuit, the load equation can be written as

$$V_{CE} = V_{CC} - I_C R_C$$

This equation involves two unknown V_{CE} and I_C and therefore can not be solved. To solve this equation output characteristic (I_C vs V_{CE}) is used.

The load equation is the equation of a straight line and given by two points:

$I_C = 0,$ $V_{CE} = V_{CC}$

& $V_{CE} = 0,$ $I_C = V_{CC} / R_C$

The intersection of this line which is also called dc load line and the characteristic gives the operating point Q as shown in the figure below.

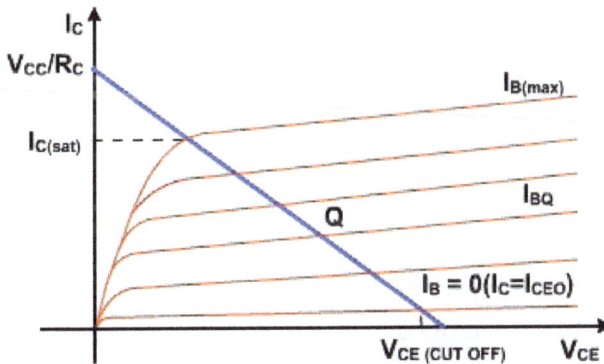

The point at which the load line intersects with $I_B = 0$ characteristic is known as cut off point. At this point base current is zero and collector current is almost negligibly small.

At cut off the emitter diode comes out of forward bias and normal transistor action is lost. To a close approximation.

V_{CE} (cut off) $\approx V_{CC}$ (approximately).

The intersection of the load line and $I_B = I_{B(max)}$ characteristic is known as saturation point . At this point $I_B = I_{B(max)}$, $I_C = I_{C(sat)}$. At this point collector diodes comes out of reverse bias and again transistor action is lost. To a close approximation,

$I_{C(sat)} \approx V_{CC} / R_C$ (approximately).

The $I_{B(sat)}$ is the minimum current required to operate the transistor in saturation region. If the I_B is less than $I_{B(sat)}$, the transistor will operate in active region. If $I_B > I_{B(sat)}$ it always operates in saturation region.

If the transistor operates at saturation or cut off points and no where else then it is operating as a switch is shown in the figure below.

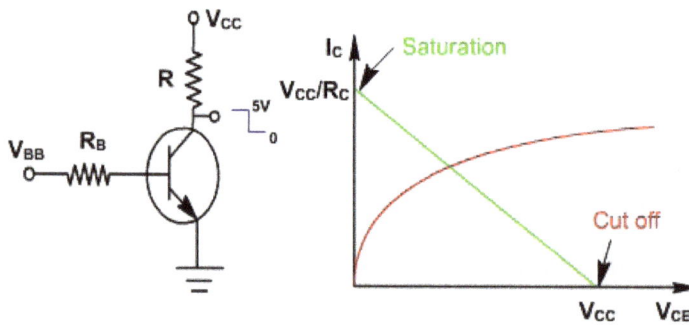

$$V_{BB} = I_B R_B + V_{BE}$$

$$I_B = (V_{BB} - V_{BE}) / R_B$$

If $I_B > I_{B(sat)}$, then it operates at saturation, If $I_B = 0$, then it operates at cut off.

If a transistor is operating as an amplifier then Q point must be selected carefully. Although we can select the operating point any where in the active region by choosing different values of R_B & R_C but the various transistor ratings such as maximum collector dissipation $P_{C(max)}$ maximum collector voltage $V_{C(max)}$ and $I_{C(max)}$ & $V_{BE(max)}$ limit the operating range.

Once the Q point is established an ac input is connected. Due to this the ac source the base current varies. As a result of this collector current and collector voltage also varies and the amplified output is obtained.

If the Q-point is not selected properly then the output waveform will not be exactly the input waveform. i.e. It may be clipped from one side or both sides or it may be distorted one.

Example-1

Find the transistor current in the circuit shown in the figure below, if I_{CO}= 20nA, β =100.

Solution

For the base circuit, $5 = 200 \times IB + 0.7$

Therefore, $I_B = \dfrac{5-0.7}{200k} = 0.0215mA$

Since $I_{CO} \ll I_B$, therefore, $I_C = \beta I_B = 2.15$ mA

From the collector circuit, $V_{CE} = 10 - 3 \times 2.15 = 3.55$ V

Since, $V_{CE} = V_{CB} + V_{BE}$

Thus, $V_{CB} = 3.55 - 0.7 = 2.55$ V

Therefore, collector junction is reverse biased and transistor is operating in its active region.

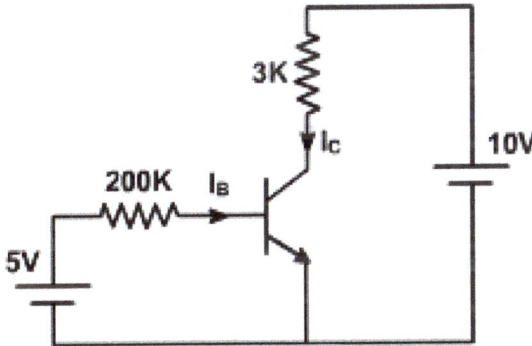

Example-2

If a resistor of 2K is connected in series with emitter in the circuit as shown in the figure below, find the currents. Given I_{CO}= 20 nA, β =100.

Solution

$I_E = I_B + I_C = I_B + 100 I_B = 101 I_B$

For the base circuit, $5 = 200 \times I_B + 0.7 + 2k \times 101 I_B$

Therefore, $I_B = \dfrac{5-0.7}{402k} = 0.0107mA$

Since $I_{CO} \ll I_B$, therefore, $I_C = \beta I_B = 1.07$ mA

From the collector circuit, $V_{CB} = 10 - 3 \times 1.07 - 0.7 - 2 \times 101 \times 0.0107 = 3.93$ V

Therefore collector junction is reverse biased and transistor is operating in its active region.

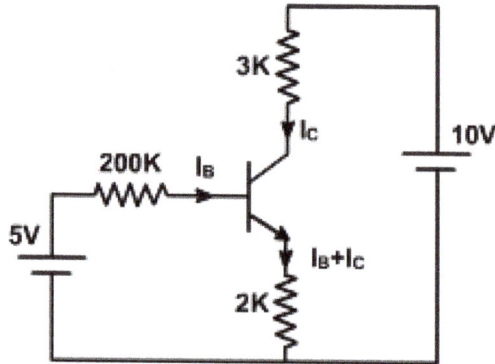

Example-3

Repeat the example-1 if R_B is replaced by 50k.

Solution:

The circuit is shown in the figure below.

Since the base resistance is reduced, the base current must have increased and there is a possibility that the transistor has entered into saturation region.

Assuming transistor is operating in its saturation region,

$$V_{BE\ (sat)} = 0.8 \text{ V and } V_{CE\ (sat)} = 0.2V$$

Therefore, $I_B = \dfrac{5 - 0.8}{50k} = 0.0840 \text{mA}$

and $\qquad I_C = \dfrac{10 - 0.2}{3k} = 3.267 \text{mA}$

The minimum base current required for operating the transistor in saturation region is

$$I_{B(min)} = \dfrac{I_C}{\beta} = 0.03267 \text{mA}$$

Since $I_B > I_{B(min)}$, therefore, transistor is operating in its saturation region.

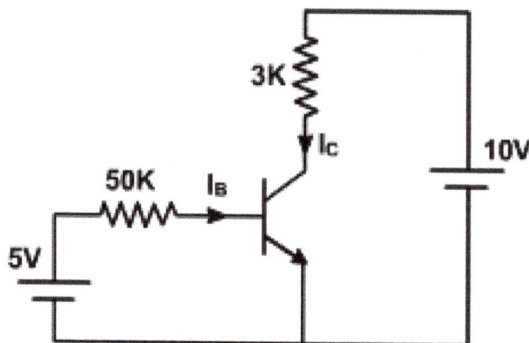

Example-4

Repeat the example-2 if R_B is replaced by 50k.

Solution:

The circuit is shown in the figure below.

Since the base resistance is reduced, the base current must have increased and there is a possibility that the transistor has entered into saturation region.

Assuming transistor is operating in its saturation region,

$$5 = 50I_B + 0.8 + 2 \times (I_B + I_C)$$
$$10 = 3I_C + 0.2 + 2 \times (I_B + I_C)$$

Solving these equations, we get,

I_C = 1.96mA and I_B = 0.0035mA

The minimum base current required for operating the transistor in saturation region is

$$I_{B(min)} = \frac{I_C}{\beta} = 0.0196mA$$

Since $I\,B < I\,B(min)$, therefore, transistor is operating in its active region and not in saturation. The base and the collector currents can be recalculated assuming the transistor to be in active region.

For the base circuit, $5 = 50 \times I_B + 0.7 + 2k \times 101\,I_B$

Therefore, $I_B = \dfrac{5-0.7}{252k} = 0.0171mA$

I_C = 1.71mA

From the collector circuit, $V_{CB} = 10 - 3 \times 1.71 - 0.7 - 2 \times 101 \times 0.0171 = 0.716$ V

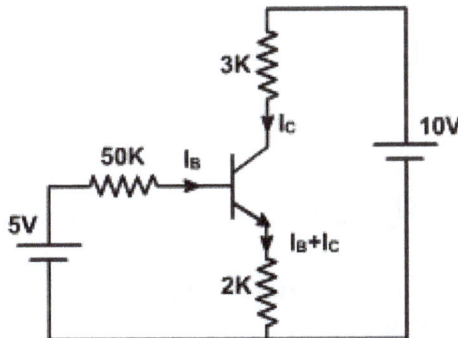

Example-1

Determine the Q-point for the CE amplifier given in the figure below, if $R_1 = 1.5K\ \Omega$ and $R_s = 7K\ \Omega$. A 2N3904 transistor is used with $\beta = 180$, $R_E = 100\Omega$ and $R_C = R_{load} = 1K\ \Omega$. Also determine the $P_{out}(ac)$ and the dc power delivered to the circuit by the source.

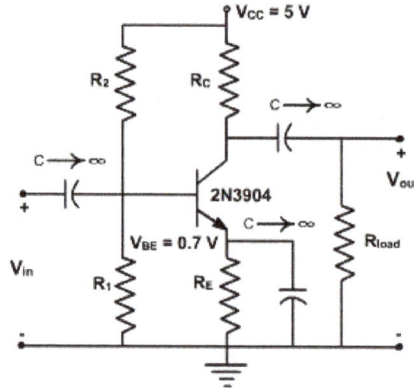

Solution

We first obtain the Thevenin equivalent.

$$V_{BB} = \frac{R_1}{R_1 + R_2} V_{CC} = \frac{1500}{1500 + 7000} .5 = 0.882V$$

and

$$R_B = \frac{R_1 R_2}{R_1 + R_2} = 1.24k\Omega$$

$$I_{CQ} = \frac{V_{BB} - V_{BE}}{R_B/\beta + R_E} = \frac{0.882 - 0.7}{1240/180 + 100} 1.70\,mA$$

Note that this is not a desirable Q-point location since V_{BB} is very close to V_{BE}. Variation in V_{BE} therefore significantly change I_C. We find $R_{ac} = R_C \,||\, R_{load} = 500$ W and $R_{dc} = R_C + R_E = 1.1K\Omega$. The value of V_{CE} representing the quiescent value associated with I_{CQ} is found as follows,

$$V_{CEQ} = V_{CC} - I_{CQ}R_{dc} = 5 - (1.70 \times 10^{-3})(1.1 \times 10^{-3}) = 3.13V$$

Then

$$V_{CC} = V_{CEQ} + I_{CQ}R_{ac} = 3.13 + (1.7 \times 10^{-3})(500) = 3.98V$$

Since the Q-point is on the lower half of the ac load line, the maximum possible symmetrical output voltage swing is

$$2(I_{CQ}-0)(R_C \| R_{load}) = 2(1.70\times10^{-3})(500) = 1.70\ V_{peak-peak}$$

The ac power output can be calculated as

$$P_{out}(ac) = \frac{1}{2}i_{load}^2 R_{load} = \frac{1}{2}\left(1.70\times10^{-3}\times\frac{1000}{2000}\right)^2 \times 1000 = 0.361mW$$

The power drawn from the dc source is given by

$$P_{V_{CC}}(dc) = I_{CQ}V_{CC} + \frac{V_{CC}^2}{R_1+R_2} = 11.4mW$$

The power loss in the transistor is given by

$$P_{transistor} = V_{CEQ}I_{CQ} = 3.13V \times 1.70mA = 5.32mW$$

The Q-point in this example is not in the middle of the load line so that output swing is not as great as possible. However, if the input signal is small and maximum output is not required, a small I_C can be used to reduce the power dissipated in the circuit.

Moving Ground Around:

Ground is a reference point that can be moved around. e.g. consider a collector feed-back bias circuit. The various stages of moving ground are shown in the figure below.

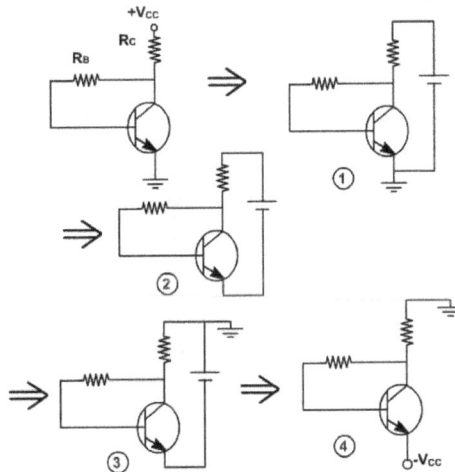

Biasing a pnp Transistor:

The biasing of pnp transistor is done similar to npn transistor except that supply is of opposite polarity The various biasing circuits of pnp transistor are shown in the figure below.

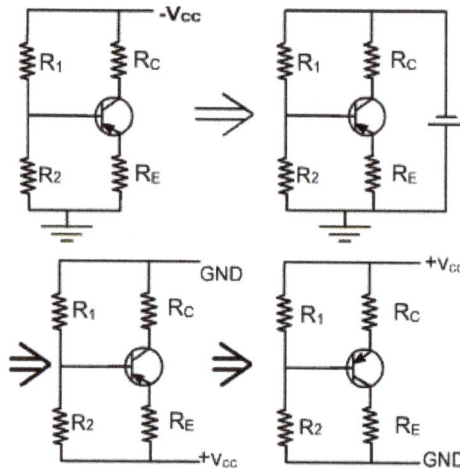

Example 2

For the circuit shown in the figure below, calculate I_C and V_{CE}

Solution

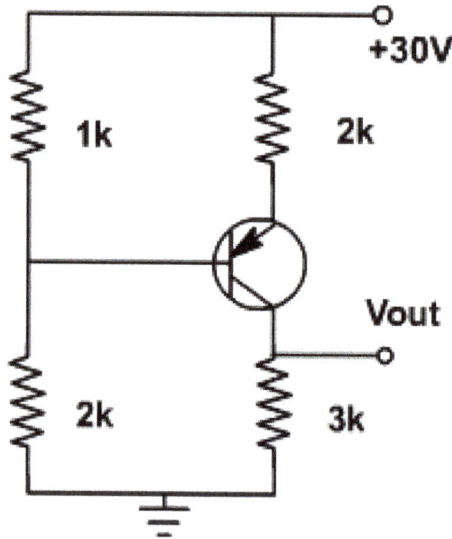

Voltage across 1K ohm resistor $= \dfrac{1}{3} \times 30 = 10V$

Therefore,

$$I_C \approx I_E = \frac{10-0.7}{2K} = \frac{9.3}{2K} = 4.65mA$$

Therefore, $V_C = 465 \times 3K = 13.95V$

Unijunction Transistor

A unijunction transistor (UJT) is a three-lead electronic semiconductor device with only one junction that acts exclusively as an electrically controlled switch.

The UJT is not used as a linear amplifier. It is used in free-running oscillators, synchronized or triggered oscillators, and pulse generation circuits at low to moderate frequencies (hundreds of kilohertz). It is widely used in the triggering circuits for silicon controlled rectifiers. The low cost per unit, combined with its unique characteristic, have warranted its use in a wide variety of applications like oscillators, pulse generators, saw-tooth generators, triggering circuits, phase control, timing circuits, and voltage- or current-regulated supplies. The original unijunction transistor types are now considered obsolete, but a later multi-layer device, the programmable unijunction transistor (PUT), is still widely available.

Types

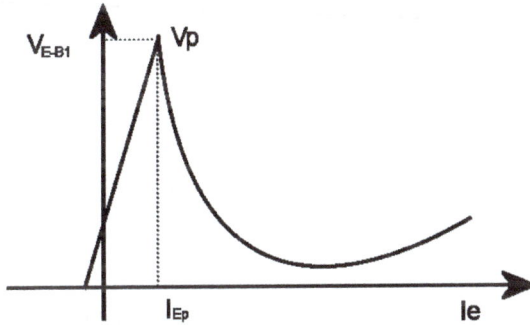

Graph of UJT characteristic curve, emitter-base1 voltage as a function of emitter current, showing current-controlled negative resistance (downward-sloping region)

There are three types of unijunction transistor:

1. The original unijunction transistor, or UJT, is a simple device that is essentially a bar of n-type semiconductor material into which p-type material has been diffused somewhere along its length, fixing the device parameter η (the "intrinsic stand-off ratio"). The 2N2646 model is the most commonly used version of the UJT.

2. The complementary unijunction transistor, or CUJT, is a bar of p-type semiconductor material into which n-type material has been diffused somewhere along its length, defining the device parameter η. The 2N6114 model is one version of the CUJT.

3. The programmable unijunction transistor, or PUT, is a multi-junction device that, with two external resistors, displays similar characteristics to the UJT. It is a close cousin to the thyristor and like the thyristor consists of four p-n layers. It has an anode and a cathode connected to the first and the last layer, and a

gate connected to one of the inner layers. PUTs are not directly interchangeable with conventional UJTs but perform a similar function. In a proper circuit configuration with two "programming" resistors for setting the parameter η, they behave like a conventional UJT. The 2N6027, 2N6028 and BRY39 models are examples of such devices.

Applications

Unijunction transistor circuits were popular in hobbyist electronics circuits in the 1960s and 1970s because they allowed simple oscillators to be built using just one active device. For example, they were used for relaxation oscillators in variable-rate strobe lights. Later, as integrated circuits became more popular, oscillators such as the 555 timer IC became more commonly used.

In addition to its use as the active device in relaxation oscillators, one of the most important applications of UJTs or PUTs is to trigger thyristors (silicon controlled rectifiers (SCR), TRIAC, etc.). A DC voltage can be used to control a UJT or PUT circuit such that the "on-period" increases with an increase in the DC control voltage. This application is important for large AC current control.

UJTs can also be used to measure magnetic flux. The hall effect modulates the voltage at the PN junction. This affects the frequency of UJT relaxation oscillators. This only works with UJTs. PUTs do not exhibit this phenomenon.

Construction

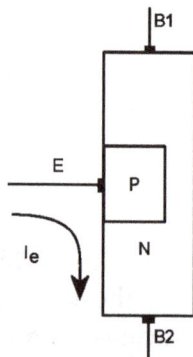

Structure of a p-type UJT

The UJT has three terminals: an emitter (E) and two bases (B_1 and B_2) and so is sometimes known a "double-base diode". The base is formed by a lightly doped n-type bar of silicon. Two ohmic contacts B_1 and B_2 are attached at its ends. The emitter is of p-type and is heavily doped; this single PN junction gives the device its name. The resistance between B1 and B2 when the emitter is open-circuit is called *interbase resistance*. The emitter junction is usually located closer to base-2 (B2) than base-1 (B1) so that the device is not symmetrical, because a symmetrical unit does not provide optimum electrical characteristics for most of the applications.

If no potential difference exists between its emitter and either of its base leads, there is an extremely small current (flow of charge) from B_1 to B_2. On the other hand, if an adequately large voltage relative to its base leads, known as the *trigger voltage*, is applied to its emitter, then a very large current from its emitter joins the current from B_1 to B_2, which creates a larger B_2 output current.

The schematic diagram symbol for a unijunction transistor represents the emitter lead with an arrow, showing the direction of conventional current when the emitter-base junction is conducting a current. A complementary UJT uses a p-type base and an n-type emitter, and operates the same as the n-type base device but with all voltage polarities reversed.

The structure of a UJT is similar to that of an N-channel JFET, but p-type (gate) material surrounds the N-type (channel) material in a JFET, and the gate surface is larger than the emitter junction of UJT. A UJT is operated with the emitter junction forward-biased while the JFET is normally operated with the gate junction reverse-biased. It is a current-controlled negative resistance device.

Device Operation

The device has a unique characteristic that when it is triggered, its emitter current increases regeneratively until it is restricted by emitter power supply. It exhibits a negative resistance characteristic and so it can be employed as an oscillator.

The UJT is biased with a positive voltage between the two bases. This causes a potential drop along the length of the device. When the emitter voltage is driven approximately one diode voltage above the voltage at the point where the P diffusion (emitter) is, current will begin to flow from the emitter into the base region. Because the base region is very lightly doped, the additional current (actually charges in the base region) causes conductivity modulation which reduces the resistance of the portion of the base between the emitter junction and the B2 terminal. This reduction in resistance means that the emitter junction is more forward biased, and so even more current is injected. Overall, the effect is a negative resistance at the emitter terminal. This is what makes the UJT useful, especially in simple oscillator circuits.

Invention

The unijunction transistor was invented as a byproduct of research on germanium tetrode transistors at General Electric. It was patented in 1953. Commercially, silicon devices were manufactured.

The UJT as the name implies, is characterized by a single pn junction. It exhibits negative resistance characteristic that makes it useful in oscillator circuits.

The symbol for UJT is shown in the left figure. The UJT is having three terminals base1 (B1), base2 (B2) and emitter (E). The UJT is made up of an N-type silicon bar which acts as the base as shown in the right figure. It is very lightly doped. A P-type impurity is introduced into the base, producing a single PN junction called emitter. The PN junction exhibits the properties of a conventional diode.

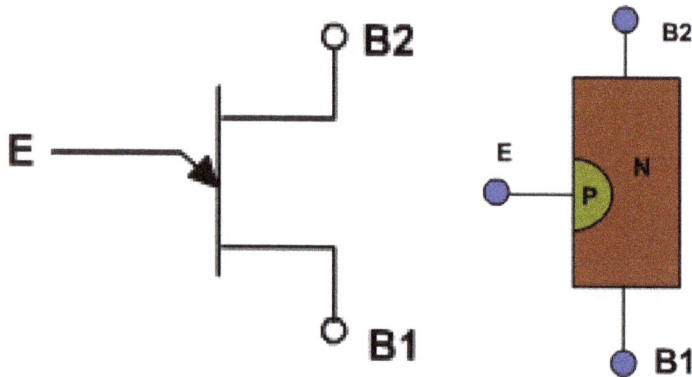

A complementary UJT is formed by a P-type base and N-type emitter. Except for the polarity of voltage and current the characteristic is similar to those of a conventional UJT.

A simplified equivalent circuit for the UJT is shown in the following figure. V_{BB} is a source of biasing voltage connected between B2 and B1. When the emitter is open, the total resistance from B2 to B1 is simply the resistance of the silicon bar, this is known as the inter base resistance R_{BB}. Since the N-channel is lightly doped, therefore R_{BB} is relatively high, typically 5 to 10K ohm. R_{B2} is the resistance between B2 and point 'a', while R_{B1} is the resistance from point 'a' to B1, therefore the interbase resistance R_{BB} is

$$R_{BB} = R_{B1} + R_{B2}$$

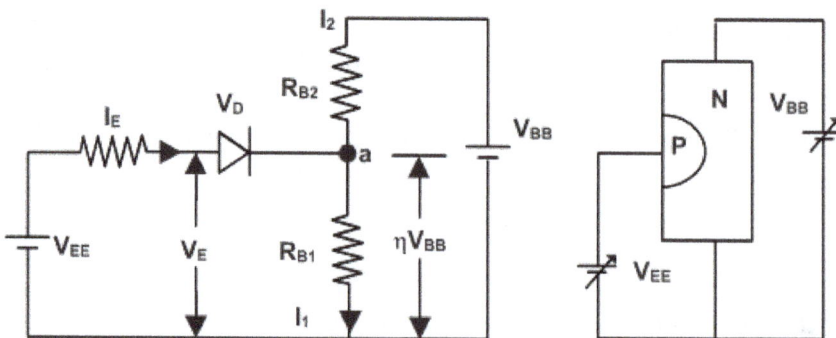

The diode accounts for the rectifying properties of the PN junction. V_D is the diode's threshold voltage. With the emitter open, $I_E = 0$, and $I_1 = I_2$. The interbase current is given by

$$I_1 = I_2 = V_{BB} / R_{BB}.$$

Part of V_{BB} is dropped across R_{B2} while the rest of voltage is dropped across R_{B1}. The voltage across R_{B1} is

$$V_a = V_{BB} * (R_{B1}) / (R_{B1} + R_{B2})$$

The ratio $R_{B1} / (R_{B1} + R_{B2})$ is called intrinsic standoff ratio

$$\eta = R_{B1} / (R_{B1} + R_{B2}) \text{ i.e. } V_a = \eta V_{BB}.$$

The ratio η is a property of UJT and it is always less than one and usually between 0.4 and 0.85. As long as $I_B = 0$, the circuit of behaves as a voltage divider.

Assume now that v_E is gradually increased from zero using an emitter supply V_{EE}. The diode remains reverse biased till v_E voltage is less than ηV_{BB} and no emitter current flows except leakage current. The emitter diode will be reversed biased.

When $v_E = V_D + \eta V_{BB}$, then appreciable emitter current begins to flow where V_D is the diode›s threshold voltage. The value of v_E that causes, the diode to start conducting is called the peak point voltage and the current is called peak point current I_P.

$$V_P = V_D + \eta V_{BB}.$$

The graph below shows the relationship between the emitter voltage and current. v_E is plotted on the vertical axis and I_E is plotted on the horizontal axis. The region from $v_E = 0$ to $v_E = V_P$ is called cut off region because no emitter current flows (except for leakage). Once v_E exceeds the peak point voltage, I_E increases, but v_E decreases. up to certain point called valley point (V_V and I_V). This is called negative resistance region. Beyond this, I_E increases with v_E this is the saturation region, which exhibits a positive resistance characteristic.

The physical process responsible for the negative resistance characteristic is called conductivity modulation. When the v_E exceeds V_P voltage, holes from P emitter are injected into N base. Since the P region is heavily doped compared with the N-region, holes are injected to the lower half of the UJT.

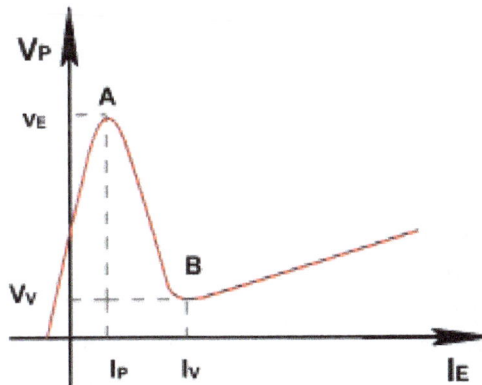

The lightly doped N region gives these holes a long lifetime. These holes move towards B1 to complete their path by re-entering at the negative terminal of V_{EE}. The large holes create a conducting path between the emitter and the lower base. These increased charge carriers represent a decrease in resistance R_{B1}, therefore can be considered as variable resistance. It decreases up to 50 ohm.

Since η is a function of R_{B1} it follows that the reduction of R_{B1} causes a corresponding reduction in intrinsic standoff ratio. Thus as I_E increases, R_{B1} decreases, η decreases, and V_a decreases. The decrease in V_a causes more emitter current to flow which causes further reduction in R_{B1}, η, and V_a. This process is regenerative and therefore V_a as well as v_E quickly drops while I_E increases. Although R_B decreases in value, but it is always positive resistance. It is only the dynamic resistance between V_v and V_p. At point B, the entire base1 region will saturate with carriers and resistance R_{B1} will not decrease any more. A further increase in I_e will be followed by a voltage rise.

The diode threshold voltage decreases with temperature and R_{BB} resistance increases with temperature because Si has positive temperature coefficient.

References

* Cubric, Marija (2007). "Analysis of the use of Wiki-based collaborations in enhancing student learning". University of Hertfordshire. Retrieved April 25, 2011

* Chelikowski, J. (2004) "Introduction: Silicon in all its Forms", p. 1 in Silicon: evolution and future of a technology. P. Siffert and E. F. Krimmel (eds.). Springer, ISBN 3-540-40546-1

* Goldman, Eric, "Wikipedia's Labor Squeeze and its Consequences", Journal on Telecommunications and High Technology Law, 8

* McFarland, Grant (2006) Microprocessor design: a practical guide from design planning to manufacturing. McGraw-Hill Professional. p. 10. ISBN 0-07-145951-0

* "Milestones:Invention of the First Transistor at Bell Telephone Laboratories, Inc., 1947". IEEE Global History Network. IEEE. Retrieved December 7, 2014

* Price, Robert W. (2004). Roadmap to Entrepreneurial Success. AMACOM Div American Mgmt Assn. p. 42. ISBN 978-0-8144-7190-6

* Noveck, Beth Simone (March 2007), "Wikipedia and the Future of Legal Education", Journal of Legal Education, 57 (1) (subscription required)

* Heywang, W. and Zaininger, K. H. (2004) "Silicon: The Semiconductor Material", p. 36 in Silicon: evolution and future of a technology. P. Siffert and E. F. Krimmel (eds.). Springer, 2004 ISBN 3-540-40546-1

* "Milestones:Invention of the First Transistor at Bell Telephone Laboratories, Inc., 1947". IEEE Global History Network. IEEE. Retrieved August 3, 2011

* Paolo Antognetti and Giuseppe Massobrio (1993). Semiconductor Device Modeling with Spice. McGraw–Hill Professional. ISBN 0-07-134955-3

* Myers, Ken S. (2008), "Wikimmunity: Fitting the Communications Decency Act to Wikipedia", Harvard Journal of Law and Technology, The Berkman Center for Internet and Society, 20: 163, SSRN 916529

- Zhong Yuan Chang, Willy M. C. Sansen, Low-Noise Wide-Band Amplifiers in Bipolar and CMOS Technologies, page 31, Springer, 1991 ISBN 0792390962

- Sanders, Robert (June 28, 2005). "Nanofluidic transistor, the basis of future chemical processors". Berkeley.edu. Retrieved June 30, 2012

- Sedra, A.S. & Smith, K.C. (2004). Microelectronic circuits (Fifth ed.). New York: Oxford University Press. p. 397 and Figure 5.17. ISBN 0-19-514251-9

- "1947: Invention of the Point-Contact Transistor - The Silicon Engine - Computer History Museum". Retrieved August 10, 2016

- Paul Horowitz and Winfield Hill (1989). The Art of Electronics (2nd ed.). Cambridge University Press. ISBN 978-0-521-37095-0

- Cunningham, Ward (November 1, 2003). "Correspondence on the Etymology of Wiki". WikiWikiWeb. Retrieved March 9, 2007

- Paul Horowitz and Winfield Hill (1989). The Art of Electronics (2nd ed.). Cambridge University Press. pp. 62–66. ISBN 978-0-521-37095-0

4

Understanding Field-Effect Transistor

A field-effect transistor (FET) is a type of transistor that controls the electrical behavior of a device through the use of an electric field. Field-effect transistors have an active channel through which charge carriers travel. Two types of FETs discussed in this chapter are junction field-effect transistor and metal oxide semiconductor field-effect transistor. This chapter is an overview of the subject matter incorporating all the major aspects of field-effect transistor.

Field-effect Transistor

The field-effect transistor (FET) is a transistor that uses an electric field to control the electrical behaviour of the device. FETs are also known as unipolar transistors since they involve single-carrier-type operation. Many different implementations of field effect transistors exist. Field effect transistors generally display very high input impedance at low frequencies. The conductivity between the drain and source terminals is controlled by an electric field in the device, which is generated by the voltage difference between the body and the gate of the device.

History

The field-effect transistor was first patented by Julius Edgar Lilienfeld in 1926 and by Oskar Heil in 1934, but practical semiconducting devices (the JFET) were developed only much later after the transistor effect was observed and explained by the team of William Shockley at Bell Labs in 1947, immediately after the 20-year patent period eventually expired. The MOSFET, which largely superseded the JFET and had a profound effect on digital electronic development, was invented by Dawon Kahng and Martin Atalla in 1959.

Basic Information

FETs can be majority-charge-carrier devices, in which the current is carried predominantly by majority carriers, or minority-charge-carrier devices, in which the current is mainly due to a flow of minority carriers. The device consists of an active channel through which charge carriers, electrons or holes, flow from the source to the drain. Source and drain terminal conductors are connected to the semiconductor through ohmic contacts. The conductivity of the channel is a function of the potential applied across the gate and source terminals.

The FET's three terminals are:

- Source (S), through which the carriers enter the channel. Conventionally, current entering the channel at S is designated by I_S.

- Drain (D), through which the carriers leave the channel. Conventionally, current entering the channel at D is designated by I_D. Drain-to-source voltage is V_{DS}.

- Gate (G), the terminal that modulates the channel conductivity. By applying voltage to G, one can control I_D.

More About Terminals

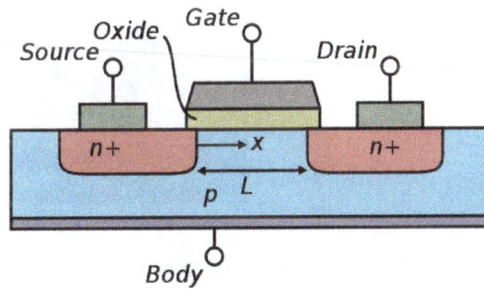

Cross section of an n-type MOSFET

All FETs have *source*, *drain*, and *gate* terminals that correspond roughly to the *emitter*, *collector*, and *base* of BJTs. Most FETs have a fourth terminal called the *body, base, bulk*, or *substrate*. This fourth terminal serves to bias the transistor into operation; it is rare to make non-trivial use of the body terminal in circuit designs, but its presence is important when setting up the physical layout of an integrated circuit. The size of the gate, length L in the diagram, is the distance between source and drain. The *width* is the extension of the transistor, in the direction perpendicular to the cross section in the diagram (i.e., into/out of the screen). Typically the width is much larger than the length of the gate. A gate length of 1 μm limits the upper frequency to about 5 GHz, 0.2 μm to about 30 GHz.

The names of the terminals refer to their functions. The gate terminal may be thought of as controlling the opening and closing of a physical gate. This gate permits electrons to flow through or blocks their passage by creating or eliminating a channel between the source and drain. Electron-flow from the source terminal towards the drain terminal is influenced by an applied voltage. The body simply refers to the bulk of the semiconductor in which the gate, source and drain lie. Usually the body terminal is connected to the highest or lowest voltage within the circuit, depending on the type of the FET. The body terminal and the source terminal are sometimes connected together since the source is often connected to the highest or lowest voltage within the circuit, although there are several uses of FETs which do not have such a configuration, such as transmission gates and cascode circuits.

Effect of Gate Voltage on Current

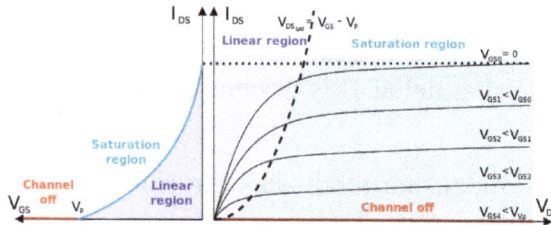

I–V characteristics and output plot of a JFET n-channel transistor.

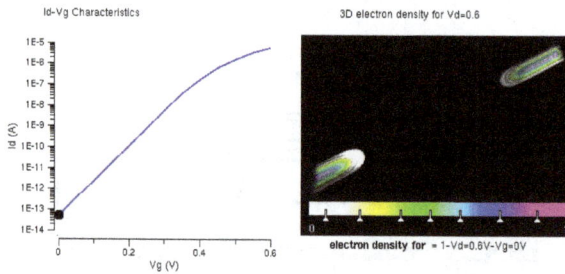

Simulation result for Right side: formation of inversion channel (electron density) and Left side: current-gate voltage curve(transfer characteristics) in a n-channel nanowire MOSFET. Note that the threshold voltage for this device lies around 0.45 V.

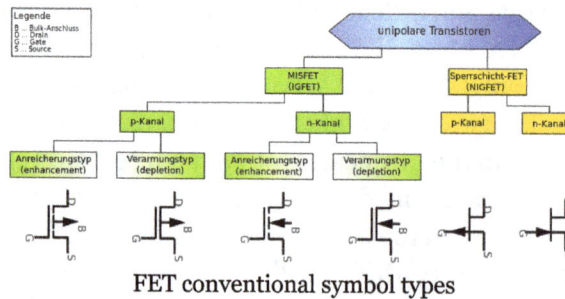

FET conventional symbol types

The FET controls the flow of electrons (or electron holes) from the source to drain by affecting the size and shape of a "conductive channel" created and influenced by voltage (or lack of voltage) applied across the gate and source terminals. (For simplicity, this discussion assumes that the body and source are connected.) This conductive channel is the "stream" through which electrons flow from source to drain.

n-channel

In an n-channel *depletion-mode* device, a negative gate-to-source voltage causes a *depletion region* to expand in width and encroach on the channel from the sides, narrowing the channel. If the active region expands to completely close the channel, the resistance of the channel from source to drain becomes large, and the FET is effectively turned off like a switch (see figure, when there is very small current). This is called *pinch-off,* and the voltage at which it occurs is called the *pinch-off voltage.* Conversely, a positive gate-to-source voltage increases the channel size and allows electrons to flow easily.

In an n-channel *enhancement-mode* device, a conductive channel does not exist naturally within the transistor, and a positive gate-to-source voltage is necessary to create one. The positive voltage attracts free-floating electrons within the body towards the gate, forming a conductive channel. But first, enough electrons must be attracted near the gate to counter the dopant ions added to the body of the FET; this forms a region with no mobile carriers called a depletion region, and the voltage at which this occurs is referred to as the *threshold voltage* of the FET. Further gate-to-source voltage increase will attract even more electrons towards the gate which are able to create a conductive channel from source to drain; this process is called *inversion.*

p-channel

In a p-channel *depletion-mode* device, a positive voltage from gate to body creates a depletion layer by forcing the positively charged holes to the gate-insulator/semiconductor interface, leaving exposed a carrier-free region of immobile, negatively charged acceptor ions.

Effect of Source/Drain Voltage on Channel

For either enhancement- or depletion-mode devices, at drain-to-source voltages much less than gate-to-source voltages, changing the gate voltage will alter the channel resistance, and drain current will be proportional to drain voltage. In this mode the FET operates like a variable resistor and the FET is said to be operating in a *linear mode* or *ohmic mode.*

If drain-to-source voltage is increased, this creates a significant asymmetrical change in the shape of the channel due to a gradient of voltage potential from source to drain. The shape of the inversion region becomes "pinched-off" near the drain end of the channel. If drain-to-source voltage is increased further, the pinch-off point of the channel begins to move away from the drain towards the source. The FET is said to be in *saturation mode*; although some authors refer to it as *active mode*, for a better analogy with bipolar transistor operating regions. The saturation mode, or the region between ohmic and saturation, is used when amplification is needed. The in-between region is sometimes considered to be part of the ohmic or linear region, even where drain current is not approximately linear with drain voltage.

Even though the conductive channel formed by gate-to-source voltage no longer connects source to drain during saturation mode, carriers are not blocked from flowing. Considering again an n-channel enhancement-mode device, a depletion region exists in the p-type body, surrounding the conductive channel and drain and source regions. The electrons which comprise the channel are free to move out of the channel through the depletion region if attracted to the drain by drain-to-source voltage. The depletion region is free of carriers and has a resistance similar to silicon. Any increase of the drain-to-source voltage will increase the distance from drain to the pinch-off point, increasing the resistance of the depletion region in proportion to the drain-to-source voltage applied. This proportional change causes the drain-to-source current to remain relatively fixed, independent of changes to the drain-to-source voltage, quite unlike its ohmic behavior in the linear mode of operation. Thus, in saturation mode, the FET behaves as a constant-current source rather than as a resistor, and can effectively be used as a voltage amplifier. In this case, the gate-to-source voltage determines the level of constant current through the channel.

Composition

FETs can be constructed from various semiconductors, with silicon being by far the most common. Most FETs are made by using conventional bulk semiconductor processing techniques, using a single crystal semiconductor wafer as the active region, or channel.

Among the more unusual body materials are amorphous silicon, polycrystalline silicon or other amorphous semiconductors in thin-film transistors or organic field-effect transistors (OFETs) that are based on organic semiconductors; often, OFET gate insulators and electrodes are made of organic materials, as well. Such FETs are manufactured using a variety of materials such as silicon carbide (SiC), gallium arsenide (GaAs), gallium nitride (GaN), and indium gallium arsenide (InGaAs).

In June 2011, IBM announced that it had successfully used graphene-based FETs in an integrated circuit. These transistors are capable of about 2.23 GHz cutoff frequency, much higher than standard silicon FETs.

Types

The channel of a FET is doped to produce either an n-type semiconductor or a p-type semiconductor. The drain and source may be doped of opposite type to the channel, in the case of enhancement mode FETs, or doped of similar type to the channel as in depletion mode FETs. Field-effect transistors are also distinguished by the method of insulation between channel and gate. Types of FETs include:

- The JFET (junction field-effect transistor) uses a reverse biased p–n junction to separate the gate from the body.

- The MOSFET (metal–oxide–semiconductor field-effect transistor) utilizes an insulator (typically SiO_2) between the gate and the body.

- The MNOS (metal–nitride–oxide–semiconductor) transistor utilizes an nitride-oxide layer insulator between the gate and the body.

- The DGMOSFET (dual-gate MOSFET) is a FET with two insulated gates.

- The DEPFET is a FET formed in a fully depleted substrate and acts as a sensor, amplifier and memory node at the same time. It can be used as an image (photon) sensor.

- The FREDFET (fast-reverse or fast-recovery epitaxial diode FET) is a specialized FET designed to provide a very fast recovery (turn-off) of the body diode.

- The HIGFET (heterostructure insulated gate field-effect transistor) is now used mainly in research.

- The MODFET (modulation-doped field-effect transistor) uses a quantum well structure formed by graded doping of the active region.

- The TFET (tunnel field-effect transistor) is based on band-to-band tunneling.

- The IGBT (insulated-gate bipolar transistor) is a device for power control. It has a structure akin to a MOSFET coupled with a bipolar-like main conduction channel. These are commonly used for the 200–3000 V drain-to-source voltage range of operation. Power MOSFETs are still the device of choice for drain-to-source voltages of 1 to 200 V.

- The HEMT (high-electron-mobility transistor), also called a HFET (heterostructure FET), can be made using bandgap engineering in a ternary semiconductor such as AlGaAs. The fully depleted wide-band-gap material forms the isolation between gate and body.

- The ISFET (ion-sensitive field-effect transistor) can be used to measure ion concentrations in a solution; when the ion concentration (such as H^+, see pH electrode) changes, the current through the transistor will change accordingly.

- The BioFET (Biologically sensitive field-effect transistor) is a class of sensors/biosensors based on ISFET technology which are utilized to detect charged molecules; when a charged molecule is present, changes in the electrostatic field at the BioFET surface result in a measurable change in current through the transistor. These include EnFETs, ImmunoFETs, GenFETs, DNAFETs, CPFETs, BeetleFETs, and FETs based on ion-channels/protein binding.

- The MESFET (metal–semiconductor field-effect transistor) substitutes the p–n junction of the JFET with a Schottky barrier; and is used in GaAs and other III-V semiconductor materials.

- The NOMFET is a nanoparticle organic memory field-effect transistor.

- The GNRFET (graphene nanoribbon field-effect transistor) uses a graphene nanoribbon for its channel.

- The VeSFET (vertical-slit field-effect transistor) is a square-shaped junctionless

FET with a narrow slit connecting the source and drain at opposite corners. Two gates occupy the other corners, and control the current through the slit.

- The CNTFET (carbon nanotube field-effect transistor).

- The OFET (organic field-effect transistor) uses an organic semiconductor in its channel.

- The DNAFET (DNA field-effect transistor) is a specialized FET that acts as a biosensor, by using a gate made of single-strand DNA molecules to detect matching DNA strands.

- The QFET (quantum field effect transistor) takes advantage of quantum tunneling to greatly increase the speed of transistor operation by eliminating the traditional transistor's area of electron conduction.

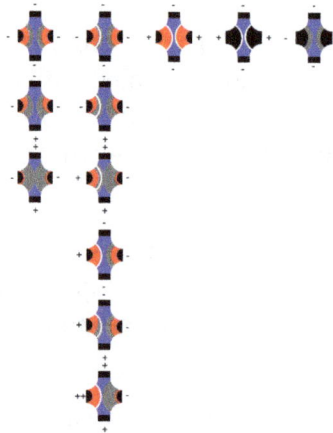

Depletion-type FETs under typical voltages: JFET, poly-silicon MOSFET, double-gate MOSFET, metal-gate MOSFET, MESFET.

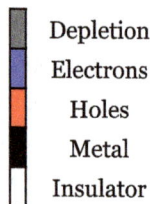

Depletion
Electrons
Holes
Metal
Insulator

Top: source, bottom: drain, left: gate, right: bulk. Voltages that lead to channel formation are not shown.

Advantages

One advantage of the FET is its high gate to main current resistance, on the order of 100 MΩ or more, thus providing a high degree of isolation between control and flow. Because base current noise will increase with shaping time, a FET typically produces less noise than a bipolar junction transistor (BJT), and is thus found in noise sensitive electronics such as tuners and low-noise amplifiers for VHF and satellite receivers. It is relatively immune to radiation. It exhibits no offset voltage at zero drain current and hence makes an excellent signal chopper. It typically has better thermal stability than a BJT. Because they are controlled by gate charge, once the gate is closed or opened, there is no additional power draw, as there would be with a bipolar junction transistor or with non-latching relays in some states. This allows extremely low-power switching, which in turn allows greater miniaturization of circuits because heat dissipation needs are reduced compared to other types of switches.

Disadvantages

It has a relatively low gain–bandwidth product compared to a BJT. The MOSFET has a drawback of being very susceptible to overload voltages, thus requiring special handling during installation. The fragile insulating layer of the MOSFET between the gate and channel makes it vulnerable to electrostatic damage or changes to threshold voltage during handling. This is not usually a problem after the device has been installed in a properly designed circuit.

FETs often have a very low 'on' resistance and have a high 'off' resistance. However the intermediate resistances are significant, and so FETs can dissipate large amounts of power while switching. Thus efficiency can put a premium on switching quickly, but this can cause transients that can excite stray inductances and generate significant voltages that can couple to the gate and cause unintentional switching. FET circuits can therefore require very careful layout and can involve trades between switching speed and power dissipation. There is also a trade-off between voltage rating and 'on' resistance, so high voltage FETs have a relatively high 'on' resistance and hence conduction losses.

Uses

The most commonly used FET is the MOSFET. The CMOS (complementary metal oxide semiconductor) process technology is the basis for modern digital integrated circuits. This process technology uses an arrangement where the (usually "enhancement-mode") p-channel MOSFET and n-channel MOSFET are connected in series such that when one is ON, the other is OFF.

In FETs, electrons can flow in either direction through the channel when operated in the linear mode. The naming convention of drain terminal and source terminal is somewhat arbitrary, as the devices are typically (but not always) built symmetrically from source to drain. This makes FETs suitable for switching analog signals between paths (multiplexing). With this concept, one can construct a solid-state mixing board, for example.

A common use of the FET is as an amplifier. For example, due to its large input resistance and low output resistance, it is effective as a buffer in common-drain (source follower) configuration.

IGBTs are used in switching internal combustion engine ignition coils, where fast switching and voltage blocking capabilities are important.

Source-gated Transistor

Source-gated transistors are more robust to manufacturing and environmental issues in large-area electronics such as display screens, but are slower in operation than FETs.

Operation of FET:

Consider a sample bar of N-type semiconductor. This is called N-channel and it is electrically equivalent to a resistance as shown in the figure below.

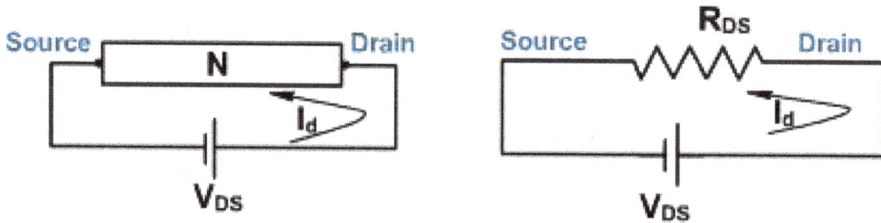

Ohmic contacts are then added on each side of the channel to bring the external connection. Thus if a voltage is applied across the bar, the current flows through the channel.

The terminal from where the majority carriers (electrons) enter the channel is called source designated by S. The terminal through which majority carriers leaves the channel is called drain and designated by D. For an N-channel device, electrons are the majority carriers. Hence the circuit behaves like a dc voltage V_{DS} applied across a resistance R_{DS}. The resulting current is the drain current I_D. If V_{DS} increases, I_D increases proportionally.

Now on both sides of the n-type bar heavily doped regions of p-type impurity have been formed by any method for creating pn junction. These impurity regions are called gates (gate1 and gate2) as shown in the figure below.

Both the gates are internally connected and they are grounded yielding zero gate source voltage (V_{GS} =0). The word gate is used because the potential applied between gate and source controls the channel width and hence the current.

As with all PN junctions, a depletion region is formed on the two sides of the reverse biased PN junction. The current carriers have diffused across the junction, leaving only uncovered positive ions on the n side and negative ions on the p side. The depletion region width increases with the magnitude of reverse bias. The conductivity of this channel is normally zero because of the unavailability of current carriers.

The potential at any point along the channel depends on the distance of that point from the drain, points close to the drain are at a higher positive potential, relative to ground, then points close to the source. Both depletion regions are therefore subject to greater reverse voltage near the drain. Therefore the depletion region width increases as we move towards drain. The flow of electrons from source to drain is now restricted to the narrow channel between the no conducting depletion regions. The width of this channel determines the resistance between drain and source.

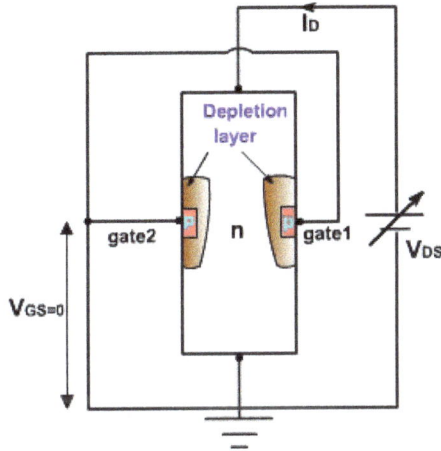

Consider now the behavior of drain current I_D vs drain source voltage V_{DS}. The gate source voltage is zero therefore $V_{GS} = 0$. Suppose that V_{DS} is gradually linearly increased linearly from 0V. I_D also increases.

Since the channel behaves as a semiconductor resistance, therefore it follows ohm's law. The region is called ohmic region, with increasing current, the ohmic voltage drop between the source and the channel region reverse biased the junction, the conducting portion of the channel begins to constrict and I_D begins to level off until a specific value of V_{DS} is reached, called the pinch of voltage V_P.

At this point further increase in V_{DS} do not produce corresponding increase in I_D. Instead, as V_{DS} increases, both depletion regions extend further into the channel, resulting in a no more cross section, and hence a higher channel resistance. Thus even though, there is more voltage, the resistance is also greater and the current remains relatively constant. This is called pinch off or saturation region. The current in this region is maximum current that FET can produce and designated by I_{DSS}. (Drain to source current with gate shorted).

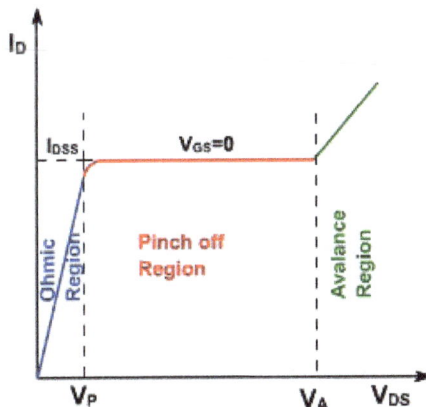

As with all pn junctions, when the reverse voltage exceeds a certain level, avalanche breakdown of pn junction occurs and I_D rises very rapidly as shown in the figure above.

Consider now an N-channel JFET with a reverse gate source voltage as shown in the figure left.

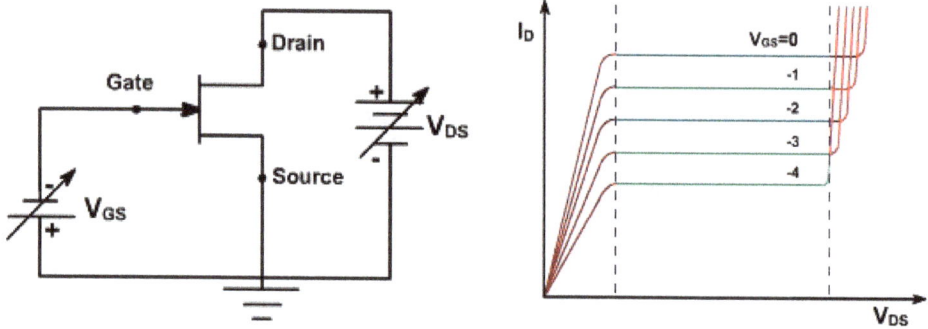

The additional reverse bias, pinch off will occur for smaller values of $|V_{DS}|$, and the maximum drain current will be smaller. A family of curves for different values of V_{GS}(negative) is shown in the figure right.

Suppose that $V_{GS}= 0$ and that due of V_{DS} at a specific point along the channel is +5V with respect to ground. Therefore reverse voltage across either p-n junction is now 5V. If V_{GS} is decreased from 0 to −1V the net reverse bias near the point is 5 - (−1) = 6V. Thus for any fixed value of V_{DS}, the channel width decreases as V_{GS} is made more negative.

Thus I_D value changes correspondingly. When the gate voltage is negative enough, the depletion layers touch each other and the conducting channel pinches off (disappears). In this case the drain current is cut off. The gate voltage that produces cut off is symbolized V_{GS}(off) . It is same as pinch off voltage.

Since the gate source junction is a reverse biased silicon diode, only a very small reverse current flows through it. Ideally gate current is zero. As a result, all the free electrons from the source go to the drain i.e. $I_D = I_S$. Because the gate draws almost negligible reverse current the input resistance is very high 10's or 100's of M ohm. Therefore where high input impedance is required, JFET is preferred over BJT. The disadvantage is less control over output current i.e. FET takes larger changes in input voltage to produce changes in output current. For this reason, JFET has less voltage gain than a bipolar amplifier.

Biasing the Field Effect Transistor

Transductance Curves:

The transductance curve of a JFET is a graph of output current (I_D) vs input voltage (V_{GS}) as shown in the figure.

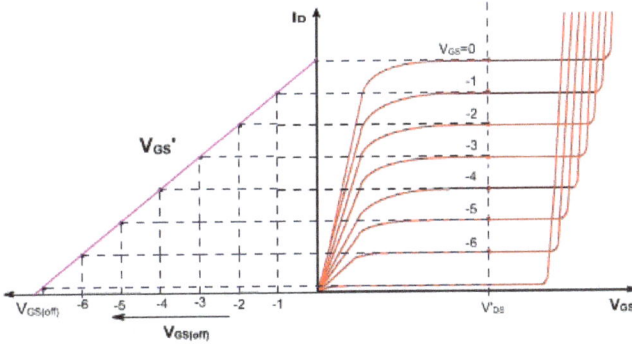

By reading the value of I_D and V_{GS} for a particular value of V_{DS}, the transductance curve can be plotted. The transductance curve is a part of parabola. It has an equation of

$$I_D = I_{DSS}\left(1 - \frac{V_{GS}}{V_{GS(off)}}\right)^2$$

Data sheet provides only I_{DSS} and V_{GS}(off) value. Using these values the transductance curve can be plotted.

Biasing the FET

The FET can be biased as an amplifier. Consider the common source drain characteristic of a JFET. For linear amplification, Q point must be selected somewhere in the saturation region. Q point is selected on the basis of ac performance i.e. gain, frequency response, noise, power, current and voltage ratings.

Gate Bias

The figure shows a simple gate bias circuit.

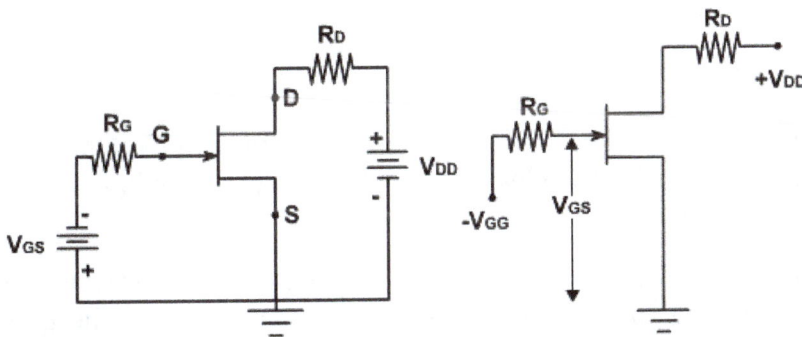

Separate V_{GS} supply is used to set up Q point. This is the worst way to select Q point. The reason is that there is considerable variation between the maximum and minimum values of FET parameters e.g.

	I_{DSS}	$V_{GS}(off)$
Minimum	4mA	-2V
Maximum	13mA	-8V

This implies that the minimum and maximum transductance curves are displaced as shown in the figure below.

Gate bias applies a fixed voltage to the gate. This fixed voltage results in a Q point that is highly sensitive to the particular JFET used. For instance, if V_{GS}= -1V the Q point may very from Q_1 to Q_2 depending upon the JFET parameter is use.

$$At\ Q_1,\ I_D = 0.016\left(1 - (1/8)\right)^2 = 12.3\ mA$$

$$At\ Q_2,\ I_D = 0.004\left(1-(1/2)\right)^2 = 1\ mA$$

The variation in drain current is very large.

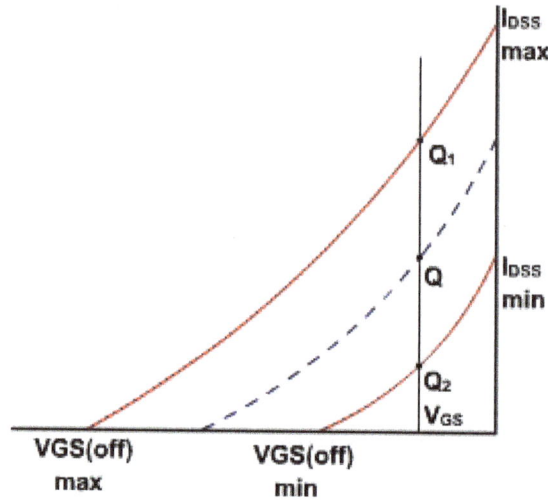

Self Bias

Figure below, shows a self bias circuit another way to bias a FET. Only a drain supply is used and no gate supply. The idea is to use the voltage across R_S to produce the gate source reverse voltage.

This is a form of a local feedback similar to that used with bipolar transistors. If drain current increases, the voltage drop across R_S increases because the I_D R_S increases. This increases the gate source reverse voltage which makes the channel narrow and reduces the drain current. The overall effect is to partially offset the original increase in drain current. Similarly, if I_D decreases, drop across R_S decreases, hence reverse bias decreases and I_D increases.

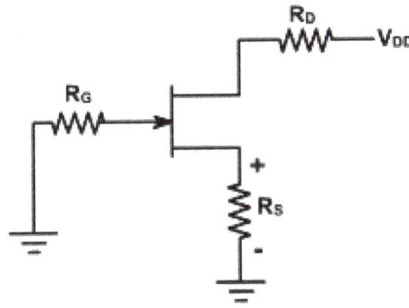

Since the gate source junction is reverse biased, negligible gate current flows through R_G and so the gate voltage with respect to ground is zero.

$$V_G = 0;$$

The source to ground voltage equals the product of the drain current and the source resistance.

$$\therefore V_S = I_D R_S.$$

The gate source voltage is the difference between the gate voltage and the source voltage.

$$V_{GS} = V_G - V_S = 0 - I_D R_S$$

$$V_{GS} = -I_D R_S.$$

This means that the gate source voltage equals the negative of the voltage across the source resistor. The greater the drain current, the more negative the gate source voltage becomes.

Rearranging the equation:

$$I_D = -V_{GS} / R_S$$

The graph of this equation is called self base line a shown in the figure below.

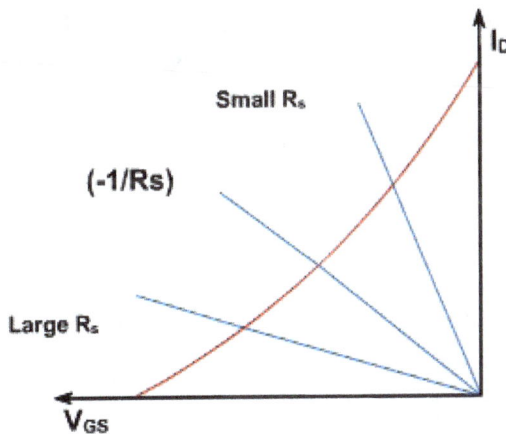

The operating point on transductance curve is the intersection of self bias line and transductance curve. The slope of the line is $(-1 / R_S)$. If the source resistance is very large ($-1 / R_S$ is small) then Q-point is far down the transductance curve and the drain current is small. When R_S is small, the Q point is far up the transductance curve and the drain current is large. In between there is an optimum value of R_S that sets up a Q point near the middle of the transductance curve.

The transductance curve varies widely for FET (because of variation in I_{DSS} and V_{GS}(off)) as shown in the figure below. The actual curve may be in between there extremes. A and B are the optimum points for the two extreme curves. To find the optimum resistance R_S, so that Q-point is correct for all the curves, A and B points are joined such that it passes through origin.

The slope of this line gives the resistance value R_S($V_{GS} = -I_D R_S$). The current I_Q is such that $I_A > I_Q > I_B$. Here A, Q and B all points are in straight line.

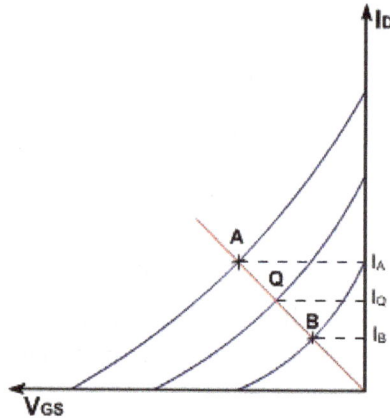

Consider the case where a line drawn to pass between points A and B does not pass through the origin. The equation $V_{GS} = - I_D R_S$ is not valid. The equation of this line is $V_{GS} = V_{GG} - I_D R_S$.

Such a bias relationship may be obtained by adding a fixed bias to the gate in addition to the source self bias as shown in the figure below.

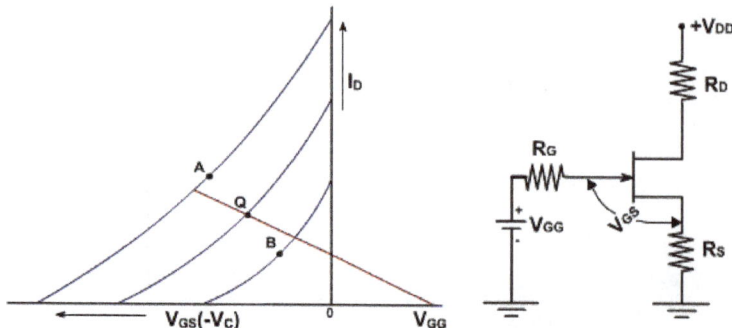

In this circuit.

$$V_{GG} = R_S I_G + V_{GS} + I_D R_S$$

Since $R_S I_G = 0$;

$$V_{GG} = V_{GS} + I_D R_S$$

or $V_{GS} = V_{GG} - I_D R_S$

Voltage Divider Bias

The biasing circuit based on single power supply is shown in the figure below. This is similar to the voltage divider bias used with a bipolar transistor.

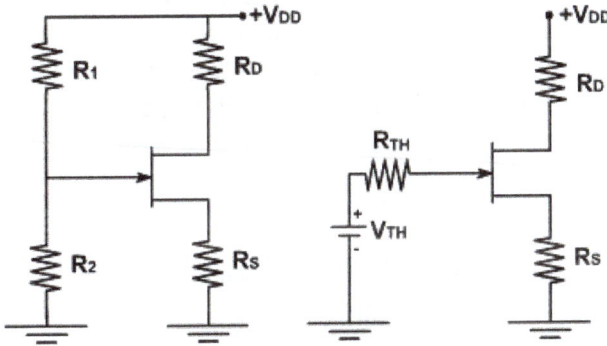

The Thevenin voltage V_{TH} applied to the gate is

$$V_{TH} = \frac{R_2}{R_1 + R_2} V_{DD}$$

The Thevenin resistance is given as

$$R_{TH} = \frac{R_2 R_1}{R_1 + R_2}$$

The gate current is assumed to be negligible. V_{TH} is the dc voltage from gate to ground.

$$V_{TH} = V_{GS} + V_S \ (neglecting \ I_G)$$
$$\therefore V_S = V_{TH} = V_{GS}$$

The drain current ID is given by

$$I_D = \frac{V_{TH} - V_{GS}}{R_S}$$

and the dc voltage from the drain to ground is $V_D = V_{DD} - I_D R_D$.

If V_{TH} is large enough to swamp out V_{GS} the drain current is approximately constant for any JFET as shown in the figure below.

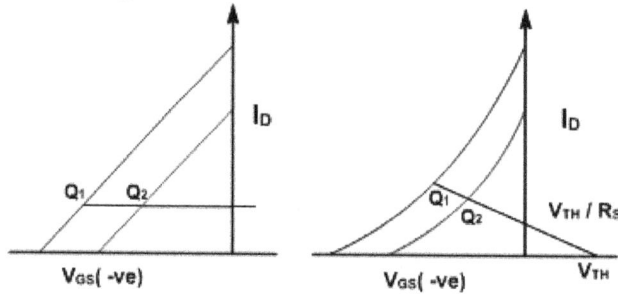

There is a problem in JFET. In a BJT, V_{BE} is approximately 0.7V, with only minor variations from one transistor to other. In a FET, V_{GS} can vary several volts from one JFET to another. It is therefore, difficult to make V_{TH} large enough to swamp out V_{GS}. For this reason, voltage divider bias is less effective with, FET than BJT. Therefore, V_{GS} is not negligible. The current increases slightly from Q2 to Q1. However, voltage divider bias maintains I_D nearly constant.

Consider a voltage divider bias circuit shown in the figure below.

$$V_{GS(min)} = -1, \quad V_{GS(max)} = -5V$$

$$V_{TH} = 15V$$

$$I_{D(min)} = \frac{15 - (-1)}{7.5K} = 2.13 \ mA$$

$$I_{D(max)} = \frac{15 - (-5)}{7.5K} = 2.67 \ mA$$

Difference in $I_{D \ (min)}$ and $I_{D \ (max)}$ is less

$$V_{D \ (max)} = 30 - 2.13 * 4.7 = 20 \ V$$

$$V_{D \ (min)} = 30 - 2.67 * 4.7 = 17.5 \ V$$

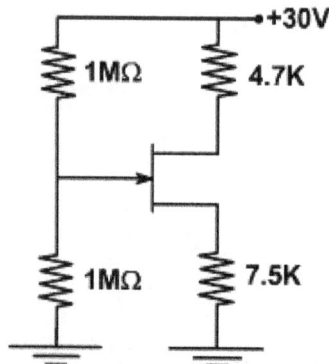

Current Source Bias:

This is another way to produce solid Q point. The aim is to produce a drain current that is independent of V_{GS}. Voltage divider bias and self bias attempt to do this by swamping out of variations in V_{GS}.

Using Two Power Supplies:

The current source bias can be used to make I_D constant the figure below.

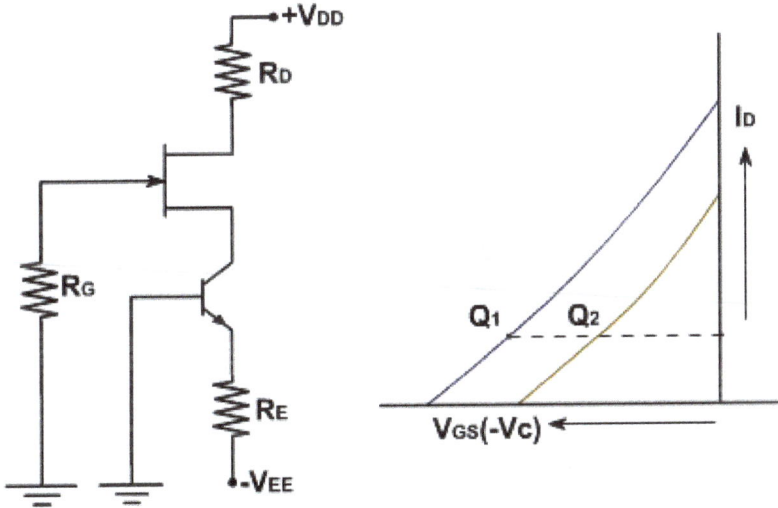

The bipolar transistor is emitter biased; its collector current is given by

$$I_c = \left(V_{EE} - V_{BE}\right)/R_E.$$

Because the bipolar transistor acts like a current source, it forces the drain current to equal the bipolar collector current.

$$I_D = I_c$$

Since I_c is constant, both Q points have the same value of drain current. The current source effectively wipes out the influence of V_{GS}. Although V_{GS} is different for each Q point, it no longer influences the value of drain current.

Using One Power Supply:

When only a positive supply is available, the circuit shown in the figure below, can be used to set up a constant drain current.

In this case, the bipolar transistor is voltage divider biased. Assuming a stiff voltage divider, the emitter and collector currents are constant for all bipolar transistors. This forces the FET drain current equal the bipolar collector current.

$$V_{TH} = \frac{R_2 V_{DD}}{R_1 + R_2}$$

$$I_E = \frac{V_{TH} - V_{BE}}{R_E}$$

since V_{TH} is constant, I_E is also constant

$I_c = I_s = I_D$ = constant

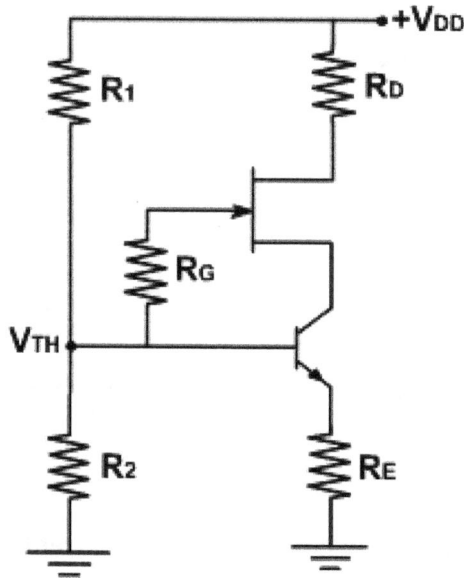

Transductance:

The transductance of a FET is defined as

$$g_m = \frac{\Delta I_D}{\Delta V_{gs}}\bigg|_{V_{DS}=0} \quad \mu A/ \text{Volts}$$

Because the changes in I_D and V_{GS} are equivalent to ac current and voltage. This equation can be written as

$$g_m = \frac{i_D}{V_{gs}}\bigg|_{V_{dS}=0}$$

The unit of g_m is mho or siemems.

Typical value of g_m is 2000 m A / V.

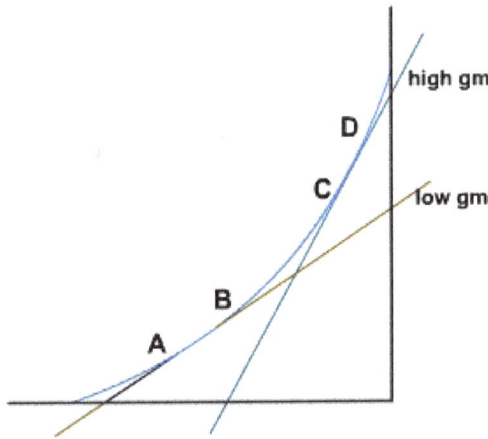

The value of g_m can be obtained from the transductance curve as shown in the figure above.

If A and B points are considered, than a change in V_{GS} produces a change in I_D. The ratio of I_D and V_{GS} is the value of g_m between A and B points. If C and D points are considered, then same change in V_{GS} produces more change in I_D. Therefore, g_m value is higher. In a nutshell, g_m tells us how much control gate voltage has over drain current. Higher the value of g_m, the more effective is gate voltage in controlling gate current. The second parameter r_d is the drain resistance.

$$r_d = \frac{v_{ds}}{i_d}\bigg|_{V_{gs}=0} \quad (r_d \text{ is negligible})$$

Junction Gate Field-effect Transistor

The junction gate field-effect transistor (JFET or JUGFET) is the simplest type of field-effect transistor. They are three-terminal semiconductor devices that can be used as electronically-controlled switches, amplifiers, or voltage-controlled resistors.

Unlike bipolar transistors, JFETs are exclusively voltage-controlled in that they do not need a biasing current. Electric charge flows through a semiconducting channel between *source* and *drain* terminals. By applying a reverse bias voltage to a *gate* terminal, the channel is "pinched", so that the electric current is impeded or switched off completely. A JFET is usually on when there is no potential difference between its gate and source terminals. If a potential difference of the proper polarity is applied between its gate and source terminals, the JFET will be more resistive to current flow, which means less current would flow in the channel between the source and drain terminals. Thus, JFETs are sometimes referred to as depletion-mode devices.

JFETs can have an n-type or p-type channel. In the n-type, if the voltage applied to the gate is less than that applied to the source, the current will be reduced (similarly in the p-type, if the voltage applied to the gate is *greater* than that applied to the source). A JFET has a large input impedance (sometimes on the order of 10^{10} ohms), which means that it has a negligible effect on external components or circuits connected to its gate.

History

A succession of FET-like devices were patented by Julius Lilienfeld in the 1920s and 1930s.However, materials science and fabrication technology would require decades of advances before FETs could actually be made. In 1947, researchers John Bardeen, Walter Houser Brattain, and William Shockley failed in their repeated attempts to make a FET. They discovered the point-contact transistor in the course of trying to diagnose the reasons for their failures. The first practical JFETs were made a decade later.

Structure

The JFET is a long channel of semiconductor material, doped to contain an abundance of positive charge carriers or holes (*p-type*), or of negative carriers or electrons (*n-type*). Ohmic contacts at each end form the source (S) and the drain (D). A pn-junction is formed on one or both sides of the channel, or surrounding it, using a region with doping opposite to that of the channel, and biased using an ohmic gate contact (G).

Function

JFET operation can be compared to that of a garden hose. The flow of water through a hose can be controlled by squeezing it to reduce the cross section and the flow of electric charge through a JFET is controlled by constricting the current-carrying channel. The current also depends on the electric field between source and drain (analogous to the difference in pressure on either end of the hose).

Constriction of the conducting channel is accomplished using the field effect: a voltage between the gate and the source is applied to reverse bias the gate-source pn-junction, thereby widening the depletion layer of this junction, encroaching upon the conducting channel and restricting its cross-sectional area. The depletion layer is so-called because it is depleted of mobile carriers and so is electrically non-conducting for practical purposes.

When the depletion layer spans the width of the conduction channel, *pinch-off* is achieved and drain-to-source conduction stops. Pinch-off occurs at a particular reverse bias (V_{GS}) of the gate-source junction. The pinch-off voltage (V_p) varies considerably, even among devices of the same type. For example, $V_{GS(off)}$ for the Temic J202 device varies from −0.8 V to −4 V. Typical values vary from −0.3 V to −10 V.

To switch off an n-channel device requires a negative gate-source voltage (V_{GS}). Conversely, to switch off a p-channel device requires positive V_{GS}.

In normal operation, the electric field developed by the gate blocks source-drain conduction to some extent.

Some JFET devices are symmetrical with respect to the source and drain.

Schematic Symbols

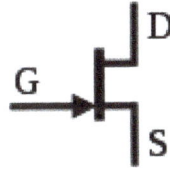

Circuit symbol for an n-Channel JFET

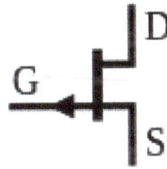

Circuit symbol for a p-Channel JFET

The JFET gate is sometimes drawn in the middle of the channel (instead of at the drain or source electrode as in these examples). This symmetry suggests that "drain" and "source" are interchangeable, so the symbol should be used only for those JFETs where they are indeed interchangeable.

Officially, the style of the symbol should show the component inside a circle (representing the envelope of a discrete device). This is true in both the US and Europe. The symbol is usually drawn without the circle when drawing schematics of integrated circuits. More recently, the symbol is often drawn without its circle even for discrete devices.

In every case the arrow head shows the polarity of the P-N junction formed between the channel and the gate. As with an ordinary diode, the arrow points from P to N, the direction of conventional current when forward-biased. An English mnemonic is that the arrow of an N-channel device "points in".

Comparison with Other Transistors

At room temperature, JFET gate current (the reverse leakage of the gate-to-channel junction) is comparable to that of a MOSFET (which has insulating oxide between gate and channel), but much less than the base current of a bipolar junction transistor. The JFET has higher gain (transconductance) than the MOSFET, as well as lower flicker noise, and is therefore used in some low-noise, high input-impedance op-amps.

Mathematical Model

The current in N-JFET due to a small voltage V_{DS} (that is, in the linear ohmic region) is given by treating the channel as a rectangular bar of material of electrical conductivity $qN_d\mu_n$:

$$I_D = \frac{bW}{L} qN_d\mu_n V_{DS}$$

where

I_D = drain–source current

b = channel thickness for a given gate voltage

W = channel width

L = channel length

q = electron charge = 1.6 x 10^{-19} C

μ_n = electron mobility

N_d = n-type doping (donor) concentration.

Linear Region

Then the drain current in the *linear region* can be expressed as:

$$I_D = \frac{bW}{L} qN_d\mu_n V_{DS} = \frac{aW}{L} qN_d\mu_n \left(1 - \sqrt{\frac{V_{GS}}{V_P}}\right) V_{DS}$$

In terms of I_{DSS}, the drain current can also be:

$$I_D = \frac{2I_{DSS}}{V_P^2} \left(V_{GS} - V_P - \frac{V_{DS}}{2}\right) V_{DS}$$

Saturation Region

The drain current in the *saturation region* is often approximated in terms of gate bias as:

$$I_{DS} = I_{DSS} \left(1 - \frac{V_{GS}}{V_P}\right)^2$$

where

I_{DSS} is the saturation current at zero gate–source voltage, i.e. the maximum current which can flow through the FET from drain to source at any (permissible) drain-to-source voltage (e. g., the I-V characteristics diagram above).

In the *saturation region*, the JFET drain current is most significantly affected by the gate–source voltage and barely affected by the drain–source voltage.

If the channel doping is uniform, such that the depletion region thickness will grow in proportion to the square root of the absolute value of the gate–source voltage, then the channel thickness b can be expressed in terms of the zero-bias channel thickness a as:

$$b = a\left(1 - \sqrt{\frac{V_{GS}}{V_P}}\right)$$

where

V_p is the pinch-off voltage, the gate–source voltage at which the channel thickness goes to zero

a is the channel thickness at zero gate–source voltage.

FET an Amplifier

Similar to Bipolar junction transistor. JFET can also be used as an amplifier. The ac equivalent circuit of a JFET is shown in the figure below.

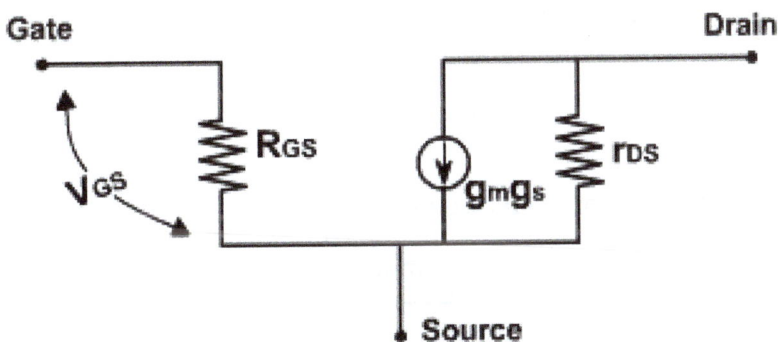

The resistance between the gate and the source R_{GS} is very high. The drain of a JFET acts like a current source with a value of $g_m V_{gs}$. This model is applicable at low frequencies.

From the ac equivalent model

$$i_d = g_m V_{gs} + \frac{V_{ds}}{r_d}$$

$$i_d = 0, \quad \frac{V_{ds}}{V_{gs}} = -g_m r_d$$

The amplification factor μ for FET is defined as

$$\mu = \frac{v_{ds}}{v_{gs}}\bigg|_{i_d=0} \qquad \therefore \mu = g_m r_d$$

When $V_{GS} = 0$, g_m has its maximum value. The maximum value is designated as g_{mo}.

Again consider the equation,

$$I_D = I_{DSS}\left[1 - \frac{V_{GS}}{V_{GS(off)}}\right]^2$$

$$g_m = \frac{\partial I_D}{\partial V_{GS}} = 2I_{DSS}\left[1 - \frac{V_{GS}}{V_{GS(off)}}\right]\left[\frac{-1}{V_{GS(off)}}\right]$$

$$gm = \frac{-2I_{DSS}}{V_{GS(off)}}\left[1 - \frac{V_{GS}}{V_{GS(off)}}\right]$$

When $\quad V_{GS} = 0, g_m = g_{mo} = \dfrac{-2I_{DSS}}{V_{GS(off)}}$

$$\therefore g_m = g_{mo} = \left[1 - \frac{V_{GS}}{V_{GS(off)}}\right]$$

As V_{GS} increases, gm decreases linearly.

$$V_{GS(off)} = \frac{-2I_{DSS}}{g_{mo}}$$

Measuring I_{DSS} and g_m, V_{GS}(off) can be determined

Figure below shows a common source amplifier.

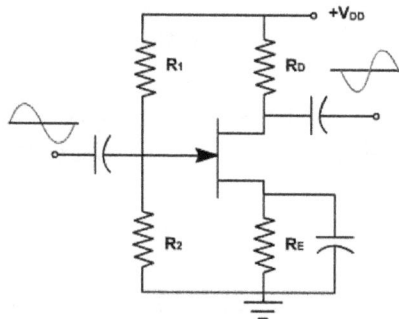

When a small ac signal is coupled into the gate it produces variations in gate source voltage. This produces a sinusoidal drain current. Since an ac current flows through the drain resistor. An amplified ac voltage is obtained at the output. An increase in gate source voltage produces more drain current, which means that the drain voltage is decreasing. Since the positive half cycle of input voltage produces the negative half cycle of output voltage, we get phase inversion in a CS amplifier.

The ac equivalent circuit is shown in the figure below.

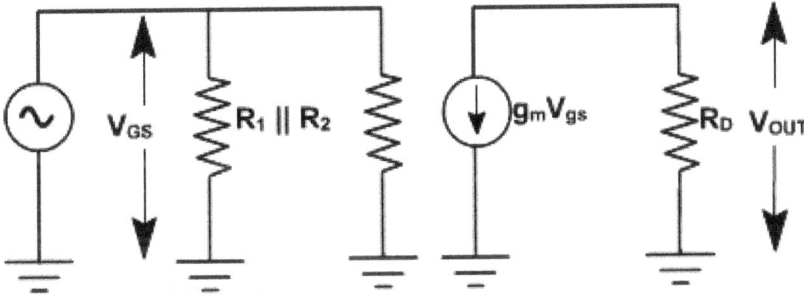

The ac output voltage is

$$V_{out} = - g_m V_{gs} R_D$$

Negative sign means phase inversion. Because the ac source is directly connected between the gate source terminals therefore ac input voltage equals

$$V_{in} = V_{gs}$$

The voltage gain is given by

$$A_V = \frac{V_{out}}{V_{in}} = -g_m \ R_D$$

$$A_V = unloaded\ voltage\ gain$$

The further simplified model of the amplifieris shown in the figure below.

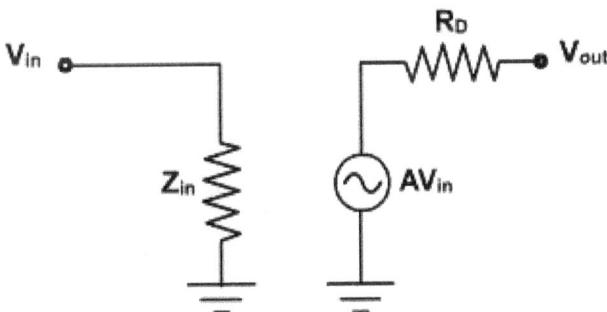

Z_{in} is the input impedance. At low frequencies, this is parallel combination of $R_1 \| R_2 \| R_{GS}$. Since R_{GS} is very large, it is parallel combination of R_1 & R_2. A V_{in} is output voltage and R_D is the output impedance.

Because of nonlinear transductance curve, a JFET distorts large signals, as shown in the figure left.

Given a sinusoidal input voltage, we get a non-sinusoidal output current in which positive half cycle is elongated and negative cycle is compressed. This type of distortion is called Square law distortion because the transductance curve is parabolic.

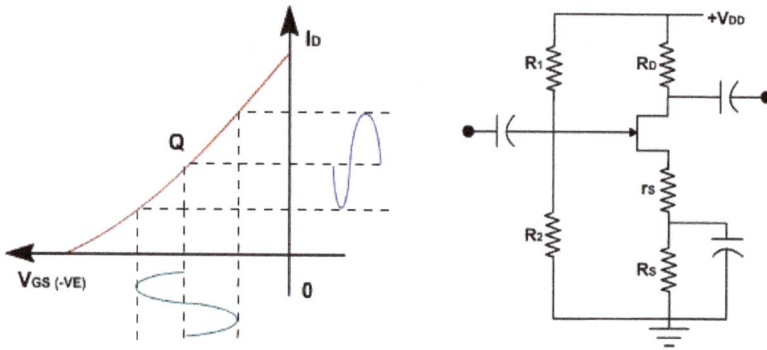

This distortion is undesirable for an amplifier. One way to minimize this is to keep the signal small. In that case a part of the curve is used and operation is approximately linear. Some times swamping resistor is used to minimize distortion and gain constant. Now the source is no longer ac ground as shown in the figure right.

The drain current through r_S produces an ac voltage between the source and ground. If r_S is large enough the local feedback can swamp out the non-linearity of the curve. Then the voltage gain approaches an ideal value of R_D / r_S.

Since R_{GS} approaches infinity therefore, all the drain current flows through r_S producing a voltage drop of $g_m V_{gs} r_S$. The ac equivalent circuit is shown in the figure below.

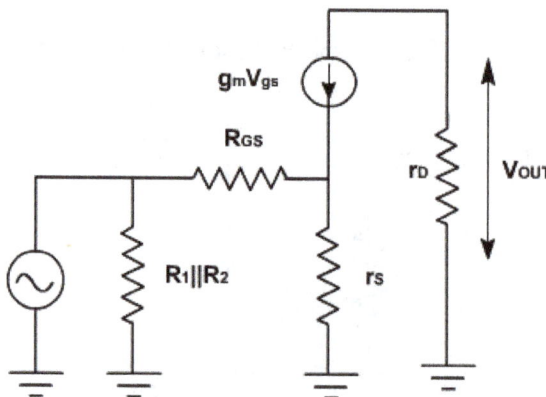

$$v_{gs} + g_m\, v_{gs}.r_s - v_{in} = 0$$

$$v_{in} = (1 + g_m r_s)\, v_{gs}$$

$$v_{out} = -g_m\, R_D\, v_{gs}$$

$$A = \frac{-g_m\, R_D}{1 + g_m\, r_s} = \frac{-R_D}{r_s + \dfrac{1}{g_m}}$$

The voltage gain reduces but voltage gain is less effective by change in g_m. r_s must be greater than $1 / g_m$ only then

$$v_{gs} = -\frac{R_D}{r_s}$$

JFET Applications

Example-1

Determine gm for an n-channel JFET with characteristic curve shown in the figure below.

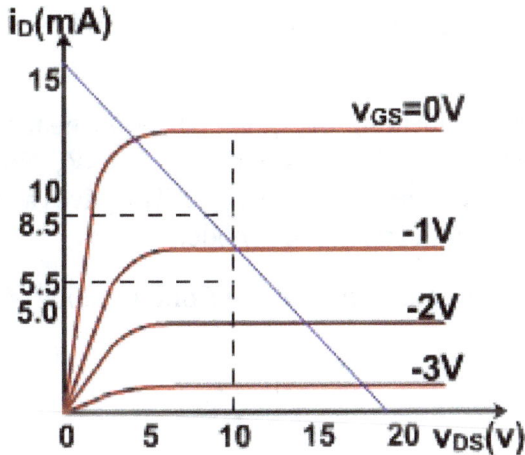

Solution

We select an operating region which is approximately in the middle of the curves; that is, between $v_{GS} = -0.8$ V and $v_{GS} = -1.2$ V; $i_D = 8.5$mA and $i_D = 5.5$ mA. Therefore, the transductance of the JFET is given by

$$g_m = \left.\frac{\Delta i_D}{\Delta v_{GS}}\right|_{V_{cs\text{-constant}}} = 7.5\ m\,\Omega^{-1}$$

Design of JFET Amplifier

To design a JFET amplifier, the Q point for the dc bias current can be determined graphically. The dc bias current at the Q point should lie between 30% and 70% of I_{DSS}. This locates the Q point in the linear region of the characteristic curves.

The relationship between i_D and v_{GS} can be plotted on a dimensionless graph (i.e., a normalized curve) as shown in the figure below.

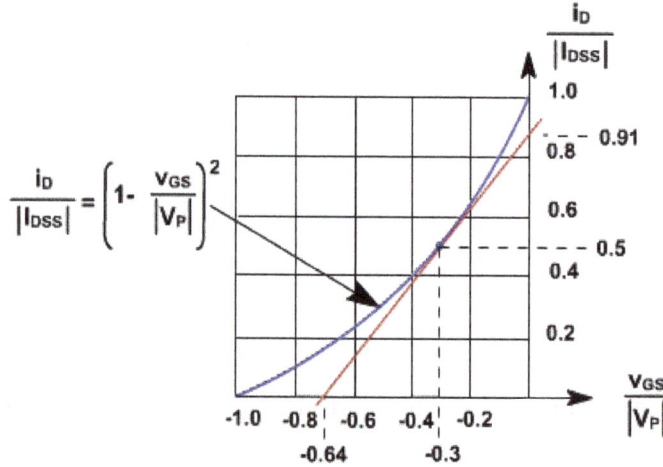

The vertical axis of this graph is i_D / I_{DSS} and the horizontal axis is v_{GS} / V_P. The slope of the curve is g_m.

A reasonable procedure for locating the quiescent point near the center of the linear operating region is to select $I_{DQ} \approx I_{DSS} / 2$ and $V_{GSQ} \approx 0.3V_P$. Note that this is near the midpoint of the curve. Next we select $v_{DS} \approx V_{DD} / 2$. This gives a wide range of values for v_{ds} that keep the transistor in the pinch –off mode.

The transductance at the Q-point can be found from the slope of the curve of figure and is given by

$$g_m = \frac{1.41\, I_{DSS}}{V_p}$$

Example-2

Determine g m for a JFET where I_{DSS} = 7 mA, V_P = -3.5 V and V_{DD} = 15V. Choose a reasonable location for the Q-point.

Solution

Let us select the Q-point as given below:

$$I_{DQ} = \frac{I_{DSS}}{2} = 3.5\ mA$$

$$V_{DSQ} = \frac{V_{DD}}{2} = 7.5\ V$$

$$V_{GSQ} = 0.3V_p = -1.05V$$

The transconductance, g_m, is found from the slope of the curve at the point $i_D / I_{DSS} = 0.5$ and $v_{GS} / V_p = 0.3$. Hence,

$$g_m = \frac{1.41\ I_{DSS}}{V_p} = 2840\ \mu\Omega^{-1}$$

JEFT as Analog Switch

JFET can be used as an analog switch as shown in the figure left. It is the major application of a JFET. The idea is to use two points on the load line: cut off and saturation. When JFET is cut off, it is like an open switch. When it is saturated, it is like a closed switch.

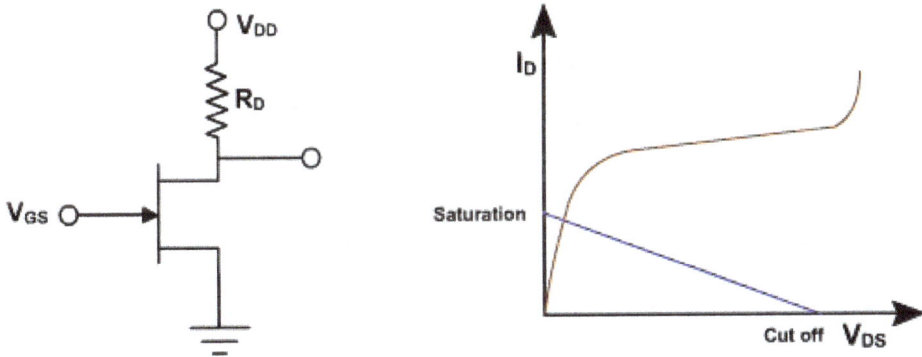

When V_{GS} =0, the JFET is saturated and operates at the upper end of the load line. When V_{GS} is equal to or more negative than V_{GS}(off) , it is cut off and operates at lower end of the load line (open and closed switch).This is shown in the figure right.

Only these two points are used for operation when used as a switch. The JFET is normally saturated well below the knee of the drain curve. For this reason the drain current is much smaller than I_{DSS}.

FET as a Shunt Switch

FET can be used as a shunt switch as shown in the figure. When V_{cont}=0, the JFT is saturated and the switch is closed When V_{cont} is more negative FET is like an open switch. The equivalent circuit is also shown in the figure below.

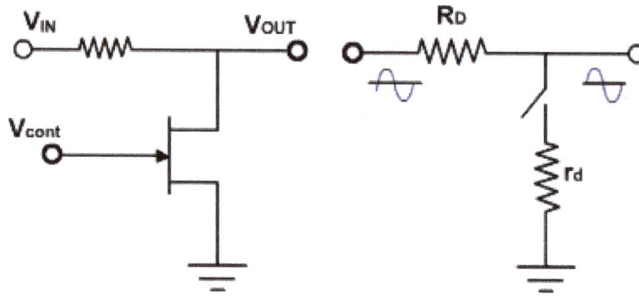

FET as a Series Switch

JFET can also be used as series switch as shown in the figure below. When control is zero, the FET is a closed switch. When V_{con} = negative, the FET is an open switch. It is better than shunt switch.

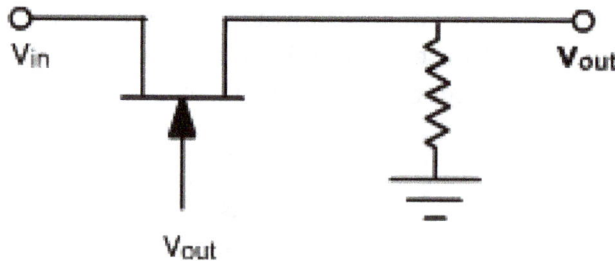

Multiplexing

One of the important application of FET is in analog multiplexer. Analog multiplexer is a circuit that selects one of the output lines as shown in the figure below. When control voltages are more negative all switches are open and output is zero. When any control voltage becomes zero the input is transmitted to the output.

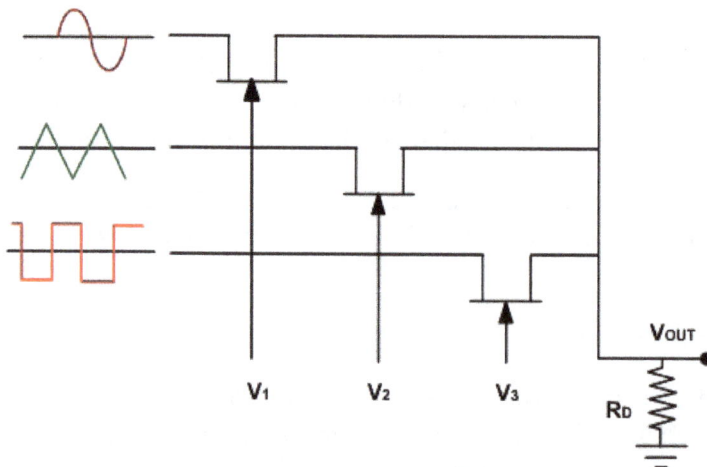

References

- C Galup-Montoro; Schneider MC (2007). MOSFET modeling for circuit analysis and design. London/Singapore: World Scientific. p. 83. ISBN 981-256-810-7

- Lin, Y.-M.; Valdes-Garcia, A.; Han, S.-J.; Farmer, D. B.; Sun, Y; Wu, Y; Dimitrakopoulos, C.; Grill, A; Avouris, P & Jenkins, K. A. (2011). "Wafer-Scale Graphene Integrated Circuit". Science. 332: 1294–1297. PMID 21659599. doi:10.1126/science.1204428

- Norbert R Malik (1995). Electronic circuits: analysis, simulation, and design. Englewood Cliffs, NJ: Prentice Hall. pp. 315–316. ISBN 0-02-374910-5

- Ionescu, A. M.; Riel, H. (2011). "Tunnel field-effect transistors as energy-efficient electronic switches". Nature. 479 (7373): 329–337. PMID 22094693. doi:10.1038/nature10679

- RR Spencer; Ghausi MS (2001). Microelectronic circuits. Upper Saddle River NJ: Pearson Education/Prentice-Hall. p. 102. ISBN 0-201-36183-3

- Poghossianb, Arshak (2002). "Recent advances in biologically sensitive field-effect transistors (BioFETs)". Analyst. 127: 1137–1151. doi:10.1039/B204444G

- Jerzy Ruzyllo (2016-09-15). Semiconductor Glossary: A Resource for Semiconductor Community. World Scientific. pp. 244–. ISBN 978-981-4749-56-5

- Sarvari H, et al. (2011). "Frequency analysis of graphene nanoribbon FET by Non-Equilibrium Green's Function in mode space". Physica E: Low-dimensional Systems and Nanostructures. 43 (8): 1509–1513. doi:10.1016/j.physe.2011.04.018

- Balbir Kumar and Shail B. Jain (2013). Electronic Devices and Circuits. PHI Learning Pvt. Ltd. pp. 342–345. ISBN 9788120348448

Permissions

Index

www.ingramcontent.com/pod-product-compliance
Lightning Source LLC
Chambersburg PA
CBHW061946190326
41458CB00009B/2799